The mu4e Reference Manual

A catalogue record for this book is available from the Hong Kong Public Libraries.

Published by Samurai Media Limited.

Email: info@samuraimedia.org

ISBN 978-988-13277-2-7

Table of Contents

Welcome to mu4e

Welcome to `mu4e` 0.9.9.6pre3!

`mu4e` (mu-for-emacs) is an e-mail client for GNU-Emacs version 24 (version 23 may works as well, but is tested as much), built on top of the `mu`[1] e-mail search engine. `mu4e` is optimized for fast handling of large amounts of e-mail.

Some of `mu4e`'s highlights:

- Fully search-based: there are no folders[2], only queries.
- Fully documented, with example configurations
- User-interface optimized for speed, with quick key strokes for common actions
- Support for non-English languages (so "angstrom" will match "Angstrm")
- Asynchronous: heavy actions don't block `emacs`[3]
- Support for crypto
- Address auto-completion based on the contacts in your messages
- Extendable with your own snippets of elisp

In this manual, we go through the installation of `mu4e`, do some basic configuration and explain its daily use. We also show you how you can customize `mu4e` for your needs.

At the end of the manual, there are some example configurations, to get you up to speed quickly: Appendix B [Example configurations], page 45. There's also an Appendix C [FAQ], page 51, which should help you with some common questions.

---

[1] http://www.djcbsoftware.nl/code/mu

[2] that is, instead of folders, you use queries that match messages in a particular folder

[3] currently, the only exception to this is *sending mail*; there are solutions for that though - see the Appendix C [FAQ], page 51

1 Introduction

1.1 Why another e-mail client?

Fair question.

I'm not sure the world needs yet another e-mail client, but perhaps *I* do! I (the author) spend a *lot* of time dealing with e-mail, both professionally and privately. Having an efficient e-mail client is essential. Since none of the existing ones worked the way I wanted, I created my own. `emacs` is an integral part of my workflow, so it made a lot of sense to use it for e-mail as well. And as I already had written an e-mail search engine (`mu`), it seemed only logical to use that as a basis.

1.2 Other mail clients

Under the hood, `mu4e` is fully search-based, similar to programs like `notmuch`[1], `md`[2] and `sup`[3]. However, `mu4e`'s user-interface is quite different. `mu4e`'s mail handling (deleting, moving etc.) is inspired by *Wanderlust*[4] (another `emacs`-based e-mail client), `mutt`[5] and `dired`.

`mu4e` tries to keep all the 'state' in your maildirs, so you can easily switch between clients, synchronize over IMAP, backup with `rsync` and so on. If you delete the database, you won't lose any information.

1.3 What mu4e does not do

There are a number of things that `mu4e` does *not* do:

- mu/mu4e do *not* get your e-mail messages from a mail server. That task is delegated to other tools, such as `offlineimap`[6], `isync`[7] or `fetchmail`[8]. As long as the messages end up in a maildir, `mu4e` and `mu` are happy to deal with them.

- `mu4e` also does *not* implement sending of messages; instead, it depends on `smptmail` (See Info file `smtpmail`, node 'Top'), which is part of `emacs`. In addition, `mu4e` piggybacks on Gnus' message editor; See Info file `message`, node 'Top'.

Thus, many of the things an e-mail client traditionally needs to do, are delegated to other tools. This leaves `mu4e` to concentrate on what it does best: quickly finding the mails you are looking for, and handle them as efficiently as possible.

---

[1] http://notmuchmail.org

[2] https://github.com/nicferrier/md

[3] http://sup.rubyforge.org/

[4] http://www.gohome.org/wl/

[5] http://www.mutt.org/

[6] http://offlineimap.org/

[7] http://isync.sourceforge.net/

[8] http://www.fetchmail.info/

1.4 Becoming a mu4e user

If `mu4e` looks like something for you, give it a shot! We've been trying hard to make it as easy as possible to set up and use; and while you can use elisp in various places to augment `mu4e`, a lot of knowledge about programming or elisp shouldn't be required. The idea is always to provide sensible defaults.

When you take `mu4e` into use, it's a good idea to subscribe to the `mu`/`mu4e`-mailing list[9].

If you have suggestions for improvements or bug reports, please use the GitHub issues list[10]. In bug reports, please clearly specify the versions of `mu`/`mu4e` and `emacs` you are using, as well as any other relevant details. Also, if it is about the behavior for specific messages, please attach the raw message (that is, the message file as it exists in your maildir); you can of course strip it of any personal informatiion.

If you are new to all this, the somewhat paternalistic *"How to ask questions the smart way"*[11] may be a good read.

---

[9] http://groups.google.com/group/mu-discuss
[10] https://github.com/djcb/mu/issues
[11] http://www.catb.org/esr/faqs/smart-questions.html

2 Getting started

In this chapter, we go through the installation of mu4e and its basic setup. After we have succeeded in Section 2.3 [Getting mail], page 6, and see Section 2.4 [Indexing your messages], page 6, we discuss Section 2.5 [Basic configuration], page 7.

After these steps, mu4e should be ready to go!

2.1 Requirements

mu/mu4e are known to work on a wide variety of Unix- and Unix-like systems, including many Linux distributions, MacOS and FreeBSD. emacs 23 or 24 (recommended) is required, as well as Xapian[1] and GMime[2].

mu has optional support for the Guile 2.x (Scheme) programming language. There are also some GUI-tools, which require GTK+ 3.x and Webkit.

If you intend to compile mu yourself, you need to have the typical development tools, such as C and C++ compilers (both gcc and clang should work), GNU Autotools and make, and the development packages for GMime, GLib and Xapian. Optionally (if you use them), you also need the development packages for GTK+, Webkit and Guile.

2.2 Installation

mu4e is part of mu - by installing the latter, the former is installed as well. Some Linux distributions provide packaged versions of mu/mu4e; if you can use those, there is no need to compile anything yourself. However, if there are no packages for your distribution, if they are outdated, or if you want to use the latest development versions, you can follow the steps below.

First, you need make sure you have the necessary dependencies; the details depend on your distribution. If you're using another distribution (or another OS), the below at least be helpful in identifying the packages to install.

We provide some instructions for Debian, Ubuntu and Fedora; if those do not apply to you, you can follow either [Building from a release tarball], page 5 or [Building from git], page 5.

2.2.1 Dependencies for Debian/Ubuntu

```
$ sudo apt-get install libgmime-2.6-dev libxapian-dev
# if libgmime-2.6-dev is not available, try libgmime-2.4-dev

# get emacs 23 or 24 if you don't have it yet
$ sudo apt-get install emacs24

# optional
$ sudo apt-get install guile-2.0-dev html2text xdg-utils

# optional: only needed for msg2pdf and mug (toy gtk+ frontend)
$ sudo apt-get install libwebkit-dev
```

---

[1] http://xapian.org/

[2] http://spruce.sourceforge.net/gmime/

2.2.2 Dependencies for Fedora

```
$ sudo yum install gmime-devel xapian-core-devel

# get emacs 23 or 24 if you don't have it yet
$ sudo yum install emacs

# optional
$ sudo yum install html2text xdg-utils

# optional: only needed for msg2pdf and mug (toy gtk+ frontend)
$ sudo yum install webkitgtk3-devel
```

2.2.3 Building from a release tarball

Using a release-tarball (as available from GoogleCode[3], installation follows the typical steps:

```
$ tar xvfz mu-<version>.tar.gz  # use the specific version
$ cd mu-<version>
# On the BSDs: use gmake instead of make
$ ./configure && make
$ sudo make install
```

Xapian, GMime and their dependencies must be installed.

2.2.4 Building from git

Alternatively, if you build from the git repository or use a tarball like the ones that github produces, the instructions are slightly different, and require you to have autotools (Auto-conf, Automake, Libtool, and friends) installed:

```
# get from git (alternatively, use a github tarball)
$ git clone git://github.com/djcb/mu.git

$ cd mu
$ autoreconf -i && ./configure && make
# On the BSDs: use gmake instead of make
$ sudo make install
```

(Xapian, GMime and their dependencies must be installed).

After this, mu and mu4e should be installed[4] on your system, and be available from the command line in emacs.

You may need to restart emacs, so it can find mu4e in its load-path. If, even after restarting, emacs cannot find mu4e, you may need to add it to your load-path explicitly; check where mu4e is installed, and add something like the following to your configuration before trying again:

```
;; the exact path may differ -- check it
(add-to-list 'load-path "/usr/local/share/emacs/site-lisp/mu4e")
```

---

[3] http://code.google.com/p/mu0/downloads/list

[4] there's a hard dependency between versions of mu4e and mu - you cannot combine different versions

2.2.5 mu4e and emacs customization

There is some support for using the `emacs` customization system in `mu4e`, but for now, we recommend setting the values manually. Please refer to Appendix B [Example configurations], page 45 for a couple of examples of this; here we go through things step-by-step.

2.3 Getting mail

In order for `mu` (and, by extension, `mu4e`) to work, you need to have your e-mail messages stored in a *maildir*[5] - a specific directory structure with one-file-per-message. If you are already using a maildir, you are lucky. If not, some setup is required:

- *Using an external IMAP or POP server* - if you are using an IMAP or POP server, you can use tools like `getmail`, `fetchmail`, `offlineimap` or `isync` to download your messages into a maildir (`~/Maildir`, often). Because it is such a common case, there is a full example of setting `mu4e` up with `offlineimap` and Gmail; see Section B.3 [Gmail configuration], page 47.

- *Using a local mail server* - if you are using a local mail-server (such as `postfix` or `qmail`), you can teach them to deliver into a maildir as well, maybe in combination with `procmail`. A bit of googling should be able to provide you with the details.

2.4 Indexing your messages

After you have succeeded in Section 2.3 [Getting mail], page 6, we need to *index* the messages. That is - we need to scan the messages in the maildir and store the information about them in a special database. We can do that from `mu4e` – Chapter 3 [Main view], page 10, but the first time, it is a good idea to run it from the command line, which makes it easier to verify that everything works correctly.

Assuming that your maildir is at `~/Maildir`, we issue the following command:

    $ mu index --maildir=~/Maildir

This should scan your `~/Maildir`[6] and fill the database, and give progress information while doing so.

The indexing process may take a few minutes the first time you do it (for thousands of e-mails); afterwards it is much faster, since `mu` only scans messages that are new or have changed. Indexing is discussed in full detail in the `mu-index` man-page.

After the indexing process has finished, you can quickly test if everything worked, by trying some command-line searches, for example

    $ mu find hello

which lists all messages that match `hello`. For more examples of searches, see Section 7.1 [Queries], page 27, or check the `mu-find` and `mu-easy` man pages. If all of this worked well, we are well on our way setting things up; the next step is to do some basic configuration for `mu4e`.

---

[5] http://en.wikipedia.org/wiki/Maildir; in this manual we use the term 'maildir' for both the standard and the hierarchy of maildirs that store your messages

[6] In most cases, you do not even need to provide the `--maildir=~/Maildir` since it is the default; see the `mu-index` man-page for details

2.5 Basic configuration

Before we can start using mu4e, we need to tell emacs to load it. So, add to your ~/.emacs (or its moral equivalent, such as ~/.emacs.d/init.el) something like:

```
(require 'mu4e)
```

If emacs complains that it cannot find mu4e, check your load-path and make sure that mu4e's installation directory is part of it. If not, you can add it:

```
(add-to-list 'load-path MU4E-PATH)
```

with MU4E-PATH replaced with the actual path.

2.6 Folders

The next step is to tell mu4e where it can find your Maildir, and some special folders. So, for example[7]:

```
;; these are actually the defaults
(setq
  mu4e-maildir         "~/Maildir"     ;; top-level Maildir
  mu4e-sent-folder    "/sent"        ;; folder for sent messages
  mu4e-drafts-folder "/drafts"      ;; unfinished messages
  mu4e-trash-folder  "/trash"       ;; trashed messages
  mu4e-refile-folder "/archive")    ;; saved messages
```

Note, mu4e-maildir takes an actual filesystem-path, the other folder names are all relative to mu4e-maildir.

2.7 Retrieval and indexing

As we have seen, we can do all of the mail retrieval *outside* of emacs/mu4e. However, you can also do it from within mu4e. For that, set the variable mu4e-get-mail-command to the program or shell command you want to use for retrieving mail. You can then retrieve your e-mail using *M-x mu4e-update-mail-and-index*, or *C-S-u* in all mu4e-views; alternatively, you can use *C-c C-u*, which may be more convenient if you use emacs in a terminal.

If you don't have a specific command for getting mail, for example because you are running your own mail-server, you can set mu4e-get-mail-command to "true", in which case mu4e won't try to get new mail, but still re-index your messages.

You can interrupt the (foreground) update process with *q*.

You can also update your mail and index periodically in the background, by setting the variable mu4e-update-interval to the number of seconds between these updates. If set to nil, it won't update at all. After you make changes to mu4e-update-interval, mu4e must be restarted before the changes take effect.

A simple setup could look something like:

```
(setq
  mu4e-get-mail-command "offlineimap"   ;; or fetchmail, or ...
```

---

[7] Note that the folders (mu4e-sent-folder, mu4e-drafts-folder, mu4e-trash-folder and mu4e-refile-folder) can also be *functions* that are evaluated at runtime. This allows for dynamically changing them depending on context. See Chapter 9 [Dynamic folders], page 35 for details.

```
mu4e-update-interval 300)                    ;; update every 5 minutes
```

A hook `mu4e-update-pre-hook` is available which is run right before starting the process, which you can for example to influence `mu4e-get-mail-command` based on the the current situation (location, time of day, ...).

It is possible to get notifications when the indexing process does any updates - for example when receiving new mail. See `mu4e-index-updated-hook` and some tips on its usage in the Appendix C [FAQ], page 51.

2.8 Sending mail

`mu4e` re-uses Gnu's `message-mode` (See Info file `message`, node 'Top') for writing mail and inherits the setup for sending mail as well.

For sending mail using SMTP, `mu4e` uses `smtpmail` (See Info file `smtpmail`, node 'Top'). This package supports many different ways to send mail; please refer to its documentation for the details.

Here, we only provide some simple examples - for more, see Appendix B [Example configurations], page 45.

A very minimal setup:

```
;; tell message-mode how to send mail
(setq message-send-mail-function 'smtpmail-send-it)
;; if our mail server lives at smtp.example.org; if you have a local
;; mail-server, simply use 'localhost' here.
(setq smtpmail-smtp-server "smtp.example.org")
```

Since `mu4e` (re)uses the same `message` mode and `smtpmail` that Gnus uses, many settings for those also apply to `mu4e`.

2.8.1 Dealing with sent messages

By default, `mu4e` puts a copy of messages you sent in the folder determined by `mu4e-sent-folder`. In some cases, this may not be what you want - for example, when using Gmail-over-IMAP, this interferes with Gmail's handling of the sent messages folder, and you may end up with duplicate messages.

You can use the variable `mu4e-sent-messages-behavior` to customize what happens with sent messages. The default is the symbol `sent` which, as mentioned, causes the message to be copied to your sent-messages folder. Other possible values are the symbols `trash` (the sent message is moved to the trash-folder (`mu4e-trash-folder`), and `delete` to simply discard the sent message altogether (so GMail can deal with it).

For Gmail-over-IMAP, you could add the following to your settings:

```
;; don't save messages to Sent Messages, Gmail/IMAP takes care of this
(setq mu4e-sent-messages-behavior 'delete)
```

And that's it! We should now be ready to go.

For more complex needs, `mu4e-sent-messages-behavior` can also be a a parameter-less function that returns one of the metioned symbols; see the built-in documentation for the variable.

2.9 Running mu4e

After following the steps in this chapter, we hopely now have a working mu4e setup. Great! In the next chapters, we walk you through the various views in mu4e.

For your orientation, the diagram below shows how the views relate to each other, and the default key-bindings to navigate between them.

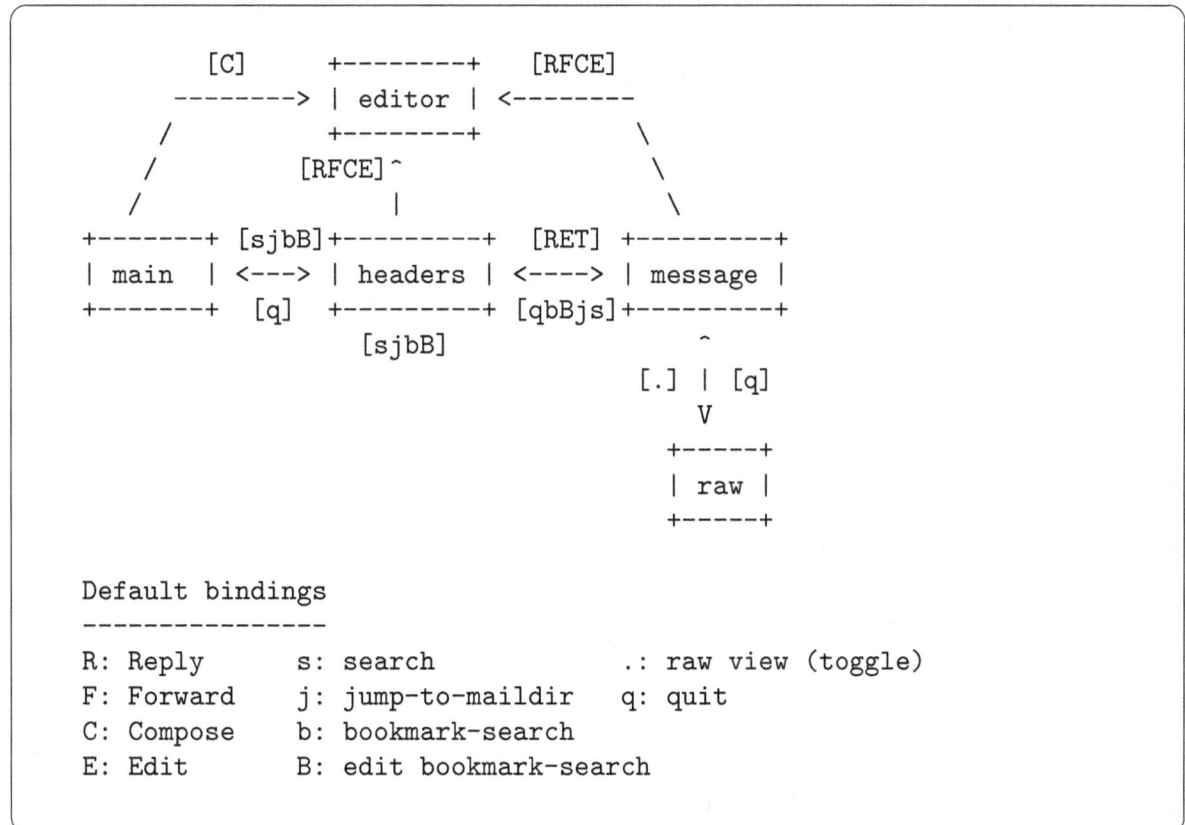

```
        [C]        +--------+    [RFCE]
     --------->  | editor |  <--------
       /          +--------+           \
      /         [RFCE]^                 \
     /             |                     \
+-------+ [sjbB]+---------+   [RET] +---------+
| main  | <---> | headers | <----> | message |
+-------+  [q]  +---------+ [qbBjs]+---------+
              [sjbB]                   ^
                               [.] | [q]
                                   V
                                +-----+
                                | raw |
                                +-----+


Default bindings
----------------

R: Reply      s: search           .: raw view (toggle)
F: Forward    j: jump-to-maildir  q: quit
C: Compose    b: bookmark-search
E: Edit       B: edit bookmark-search
```

3 The main view

After you have installed mu4e (see Chapter 2 [Getting started], page 4), you can start it with *M-x mu4e*. mu4e does some checks to ensure everything is set up correctly, and then shows you the mu4e main view. Its major mode is mu4e-main-mode.

3.1 Overview

The main view looks something like the following:

```
* mu4e - mu for emacs version x.x CG

  Basics

        * [j]ump to some maildir
        * enter a [s]earch query
        * [C]ompose a new message

  Bookmarks

        * [bu] Unread messages
        * [bt] Today's messages
        * [bw] Last 7 days
        * [bp] Messages with images
  Misc

        * [U]pdate email & database
        * toggle [m]ail sending mode (direct)
        * [f]lush queued mail

        * [A]bout mu4e
        * [H]elp
        * [q]uit mu4e
```

In the example above, you can see the letters "CG", which indicate:

- C: support for decryption of encrypted messages, and verifying signatures. See Section 5.6 [MSGV Crypto], page 21 in the Chapter 5 [Message view], page 17 for details.
- G: support for the Guile 2.0 programming language

Whether you see both, one or none of these letters depends on the way mu is built.

Let's walk through the menu.

3.2 Basic actions

First, the *Basics*:

- [j]ump to some maildir: after pressing j ("jump"), mu4e asks you for a maildir to visit. These are the maildirs you set in Section 2.5 [Basic configuration], page 7 and any of your own. If you choose o ("other") or /, you can choose from all maildirs under mu4e-maildir. After choosing a maildir, the messages in that maildir are listed, in the Chapter 4 [Headers view], page 12.

- enter a [s]earch query: after pressing s, mu4e asks you for a search query, and after entering one, shows the results in the Chapter 4 [Headers view], page 12.

- [C]ompose a new message: after pressing C, you are dropped in the Chapter 6 [Editor view], page 23 to write a new message.

3.3 Bookmarks

The next item in the Main view is *Bookmarks*. Bookmarks are predefined queries with a descriptive name and a shortcut - in the example above, we see the default bookmarks. You can view the list of messages matching a certain bookmark by pressing b followed by the bookmark's shortcut. If you'd like to edit the bookmarked query first before invoking it, use B.

Bookmarks are stored in the variable mu4e-bookmarks; you can add your own and/or replace the default ones; See Section 7.2 [Bookmarks], page 28.

3.4 Miscellaneous

Finally, there are some *Misc* (miscellaneous) actions:

- [U]pdate email & database executes the shell-command in the variable mu4e-get-mail-command, and afterwards updates the mu database; see Section 2.4 [Indexing your messages], page 6 and Section 2.3 [Getting mail], page 6 for details.

- toggle [m]ail sending mode (direct) toggles between sending mail directly, and queuing it first (for example, when you are offline), and [f]lush queued mail flushes any queued mail. This item is visible only if you have actually set up mail-queuing. Section 6.6 [Queuing mail], page 25

- [A]bout mu4e provides general information about the program

- [H]elp shows help information for this view

- Finally, [q]uit mu4e quits your mu4e-session

4 The headers view

The headers view shows the results of a query. The topline shows the names of the fields. Below that, there is a line with those fields, for each matching message, followed by a footer line. The major-mode for the headers view is `mu4e-headers-mode`.

4.1 Overview

An example headers view:

```
Date V        Flgs   From/To          List        Subject
06:32         Nu     To Edmund Dants  GstDev        + Re: Gstreamer-V4L...
15:08         Nu     Abb Busoni       GstDev          + Re: Gstreamer-V...
18:20         Nu     Pierre Morrel    GstDev           \ Re: Gstreamer...
2013-03-18    S      Jacopo           EmacsUsr     + emacs server on win...
2013-03-18    S      Mercds           EmacsUsr     \ RE: emacs server ...
2013-03-18    S      Beachamp         EmacsUsr     + Re: Copying a whole...
22:07         Nu     Albert de Moncerf EmacsUsr     \ Re: Copying a who...
2013-03-18    S      Gaspard Caderousse GstDev     | Issue with GESSimpl...
2013-03-18    Ss     Baron Danglars   GuileUsr     | Guile-SDL 0.4.2 ava...
End of search results
```

Some notes to explain what you see in the example:

- The fields shown in the headers view can be influenced by customizing the variable `mu4e-headers-fields`; see `mu4e-header-info` for the list of built-in fields. Apart from the built-in fields, you can also create custom fields using `mu4e-header-info-custom`; see Section 4.5 [HV Custom headers], page 15 for details.

- By default, the date is shown with the `:human-date` field, which shows the *time* for today's messages, and the *date* for older messages. If you want to distinguish between 'today' and 'older', you can use the `:date` field instead.

- You can customize the date and time formats with the variable `mu4e-headers-date-format` and `mu4e-headers-time-format`, respectively. In the example, we use `:human-date`, which shows when the time when the message was sent today, and the date otherwise.

- The header field used for sorting is indicated by "V" or "^"[1], corresponding to the sort order (descending or ascending, respectively). You can influence this by a mouse click, or O. Not all fields allow sorting.

- Instead of showing the `From:` and `To:` fields separately, you can use From/To (`:from-or-to` in `mu4e-headers-fields` as a more compact way to convey the most important information: it shows `From:` *except* when the e-mail was sent by the user (i.e., you) - in that case it shows `To:` (prefixed by To[2], as in the example above). To determine whether a message was sent by you, mu4e uses the variable `mu4e-user-mail-address-list`, a list of your e-mail addresses.

---

[1] or you can use little graphical triangles; see variable `mu4e-use-fancy-chars`

[2] You can customize this by changing the variable `mu4e-headers-from-or-to-prefix` (a cons cell)

- The 'List' field shows the mailing-list a message is sent to; `mu4e` tries to create a convenient shortcut for the mailing-list name; the variable `mu4e-user-mailing-lists` can be used to add your own shortcuts.

- The letters in the 'Flags' field correspond to the following: D=*draft*, F=*flagged* (i.e., 'starred'), N=*new*, P=*passed* (i.e., forwarded), R=*replied*, S=*seen*, T=*trashed*, a=*has-attachment*, x=*encrypted*, s=*signed*, u=*unread*. The tooltip for this field also contains this information.

- The subject field also indicates the discussion threads[3].

- The headers view is *automatically updated* if any changes are found during the indexing process, and if there is no current user-interaction. If you do not want such automatic updates, set `mu4e-headers-auto-update` to `nil`.

- There is a hook-function `mu4e-headers-found-hook` available which is invoked just after `mu4e` has completed showing the messages in the headers-view.

4.2 Keybindings

Using the below key bindings, you can do various things with these messages; these actions are also listed in the `Headers` menu in the `emacs` menu bar.

```
key             description
================================================================
n,p             go to next, previous message
y               select the message view (if it's visible)
RET             open the message at point in the message view


searching
---------

s               search
S               edit last query
/               narrow the search
b               search bookmark
B               edit bookmark before search
j               jump to maildir
M-left          previous query
M-right         next query


O               change sort order
P               toggle threading
Q               toggle full-search
V               toggle skip-duplicates
W               toggle include-related


marking
-------

d               mark for moving to the trash folder
```

---

[3] using Jamie Zawinski's mail threading algorithm, http://www.jwz.org/doc/threading.html

```
=              mark for removing trash flag ('untrash')
DEL,D          mark for complete deletion
m              mark for moving to another maildir folder
r              mark for refiling
+,-            mark for flagging/unflagging
?,!            mark message as unread, read

u              unmark message at point
U              unmark *all* messages

%              mark based on a regular expression
T,t            mark whole thread, subthread

<insert>       mark for 'something' (decide later)
#              resolve deferred 'something' marks

x              execute actions for the marked messages

composition
-----------
R,F,C          reply/forward/compose
E              edit (only allowed for draft messages)


misc
----
a              execute some custom action on a header
|              pipe message through shell command
C-+,C--        increase / decrease the number of headers shown
H              get help
C-S-u          update mail & reindex
q,z            leave the headers buffer
```

4.3 Marking messages

You can *mark* messages for a certain action, such as deletion or move. After one or more messages are marked, you can then execute (`mu4e-mark-execute-all`, `x`) these actions. This two-step mark-execute sequence is similar to what e.g. `dired` does. It is how `mu4e` tries to be as quick as possible, while avoiding accidents.

The mark/unmark commands support the *region* (i.e., "selection") – so, for example, if you select some messages and press `DEL`, all messages in the region are marked for deletion.

You can mark all messages that match a certain pattern with `%`. In addition, you can mark all messages in the current thread (`T`) or sub-thread (`t`).

When you do a new search or refresh the headers buffer while you still have marked messages, you are asked what to do with those marks – whether to *apply* them before leaving, or *ignore* them. This behavior can be influenced with the variable `mu4e-headers-leave-behavior`.

For more information about marking, see Chapter 8 [Marking], page 32.

4.4 Sort order and threading

By default, `mu4e` sorts messages by date, in descending order: the most recent messages are shown at the top. In addition, the messages are *threaded*, i.e., shown in the context of a discussion thread; this also affects the sort order.

The header field used for sorting is indicated by "V" or "^"[4], indicating the sort order (descending or ascending, respectively).

You can change the sort order by clicking the corresponding field with the mouse, or with *M-x mu4e-headers-change-sorting* (O); note that not all fields can be used for sorting. You can toggle threading on/off using *M-x mu4e-headers-toggle-threading* or P. For both of these functions, unless you provide a prefix argument (C-u), the current search is updated immediately using the new parameters. You can toggle full-search (Chapter 7 [Searching], page 27) using *M-x mu4e-headers-toggle-full-search* or Q.

If you want to change the defaults for these settings, you can use the variables `mu4e-headers-sortfield` and `mu4e-headers-show-threads`.

4.5 Custom headers

Sometimes the normal headers that `mu4e` offers (Date, From, To, Subject etc.) may not be enough. For these cases, `mu4e` offers *custom headers* in both the headers-view and the message-view.

You can do so by adding a description of your custom header to `mu4e-header-info-custom`, which is a list of custom headers.

Let's look at an example – suppose we want to add a custom header that shows the number of recipients for a message, i.e., the sum of the number of recipients in the To: and Cc: fields. Let's further suppose that our function takes a message-plist as its argument (Section 11.3 [Message functions], page 40).

```
(add-to-list 'mu4e-header-info-custom
  '(:recipnum .
    ( :name "Number of recipients"
      :shortname "Recip#"
      :help "Number of recipients for this message" ;; tooltip
      :function
      (lambda (msg)
(format "%d"
    (+ (length (mu4e-message-field msg :to))
       (length (mu4e-message-field msg :cc)))))))))
```

You can now add this custom header to your `mu4e-headers-fields` just like the built-in headers. After evaluating, you headers-view should now include a new header `Recip#` with the number of recipients.

Note that this function can be used in both the headers-view and the message-view; if you need something specific for one of these, you can check for the mode in your function.

---

[4] or you can use little graphical triangles; see variable `mu4e-use-fancy-chars`

4.6 Actions

`mu4e-headers-action` (`a`) lets you pick custom actions to perform on the message at point. You can specify these actions using the variable `mu4e-headers-actions`. See Chapter 10 [Actions], page 37 for the details.

mu4e defines some default actions. One of those is for *capturing* a message: `a c` 'captures' the current message. Next, when you're editing some message, you can include the previously captured message as an attachment, using `mu4e-compose-attach-captured-message`. See `mu4e-actions.el` in the `mu4e` source distribution for more example actions.

4.7 Split view

Using the *Split view*, we can see the Chapter 4 [Headers view], page 12 and the Chapter 5 [Message view], page 17 next to each other, with the message selected in the former, visible in the latter. You can influence the way the splitting is done by customizing the variable `mu4e-split-view`. Possible values are:

- `horizontal` (this is the default): display the message view below the header view. Use `mu4e-headers-visible-lines` the set the number of lines shown (default: 8).
- `vertical`: display the message view on the right side of the header view. Use `mu4e-headers-visible-columns` to set the number of visible columns (default: 30).
- anything else: don't do any splitting

Some useful key bindings in the split view:

- `C-+` and `C--`: interactively change the number of columns or headers shown
- You can change the selected window from the headers-view to the message-view and vice-versa with `mu4e-select-other-view`, bound to `y`

5 The message view

After selecting a message in the Chapter 4 [Headers view], page 12, it appears in a message view window, which shows the message headers, followed by the message body. Its major mode is `mu4e-view-mode`.

5.1 Overview

An example message view:

```
From: randy@epiphyte.com
To: julia@eruditorum.org
Subject: Re: some pics
Flags: (seen attach)
Date: Mon 19 Jan 2004 09:39:42 AM EET
Maildir: /inbox
Attachments(2): [1]DSCN4961.JPG(1.3M), [2]DSCN4962.JPG(1.4M)


Hi Julia,

Some pics from our trip to Cerin Amroth. Enjoy!

All the best,
Randy.

On Sun 21 Dec 2003 09:06:34 PM EET, Julia wrote:

[....]
```

Some notes:

- The variable `mu4e-view-fields` determines the header fields to be shown; see `mu4e-header-info` for a list of built-in fields. Apart from the built-in fields, you can also create custom fields using `mu4e-header-info-custom`; see Section 5.7 [MSGV Custom headers], page 22.

- You can set the date format with the variable `mu4e-date-format-long`.

- By default, only the names of contacts in address fields are visible (see `mu4e-view-show-addresses` to change this). You can view the e-mail addresses by clicking on the name, or pressing M-RET.

- You can compose a message for the contact at point by either clicking [mouse-2] or pressing C.

- The body text can be line-wrapped using `longlines-mode`. mu4e defines w to toggle between the wrapped and unwrapped state. If you want to do this automatically when viewing a message, invoke `longlines-mode` in your `mu4e-view-mode-hook`.

- You can hide cited parts in messages (the parts starting with ">") using `mu4e-view-hide-cited`, bound to h. If you want to do this automatically for every message, invoke the function in your `mu4e-view-mode-hook`.

- For search-related operations, see Chapter 7 [Searching], page 27.
- You can scroll down the message using SPC; if you do this at the end of a message, it automatically takes you to the next one. If you want to prevent this behavior, set `mu4e-view-scroll-to-next` to nil.

5.2 Keybindings

You can find most things you can do with this message in the *View* menu, or by using the keyboard; the default bindings are:

```
key             description
=================================================================

n,p             go to next, previous message
y               select the headers view (if it's visible)

RET             scroll down
M-RET           open URL at point / attachment at point

SPC             scroll down, if at end, move to next message

searching
---------
s               search
e               edit last query
/               narrow the search
b               search bookmark
B               edit bookmark before search
j               jump to maildir

M-left          previous query
M-right         next query

marking
-------
d               mark for moving to the trash folder
=               mark for removing trash flag ('untrash')
DEL,D           mark for complete deletion
m               mark for moving to another maildir folder
r               mark for refiling
+,-             mark for flagging/unflagging

u               unmark message at point
U               unmark *all* messages

%               mark based on a regular expression
T,t             mark whole thread, subthread

<insert>        mark for 'something' (decide later)
```

```
#               resolve deferred 'something' marks

x               execute actions for the marked messages


composition
-----------
R,F,C           reply/forward/compose
E               edit (only allowed for draft messages)


actions
-------
g               go to (visit) numbered URL (using `browse-url')
                (or: <mouse-1> or M-RET with point on url)
                C-u g visits multiple URLs
e               extract (save) attachment (asks for number)
                (or: <mouse-2> or S-RET with point on attachment)
                C-u e extracts multiple attachments
o               open attachment (asks for number)
                (or: <mouse-1> or M-RET with point on attachment)

a               execute some custom action on the message
A               execute some custom action on an attachment


misc
----
c               copy address at point (with C-u copy long version)
w               toggle line wrapping
h               toggle showing cited parts

v               show details about the cryptographic signature

.               show the raw message view. 'q' takes you back.
C-+,C--         increase / decrease the number of headers shown
H               get help
C-S-u           update mail & reindex
q,z             leave the message view
```

For the marking commands, please refer to Section 4.3 [Marking messages], page 14.

5.3 Opening and saving attachments

By default, mu4e uses the xdg-open-program[1] or (on MacOS) the open program for opening attachments. If you want to use another program, you do so by setting the MU_PLAY_PROGRAM environment variable to the program to be used.

The default directory for extracting (saving) attachments is your home directory (~/); you can change this using the variable mu4e-attachment-dir, for example:

---

[1] http://portland.freedesktop.org/wiki/

```
(setq mu4e-attachment-dir  "~/Downloads")
```

For more flexibility, `mu4e-attachment-dir` can also be a user-provided function. This function receives two parameters: the file-name and the mime-type as found in the e-mail message[2] of the attachment, either or both of which can be `nil`. For example:

```
(setq mu4e-attachment-dir
  (lambda (fname mtype)
    (cond
      ;; docfiles go to ~/Desktop
      ((and fname (string-match "\\.doc$" fname))  "~/Desktop")
      ;; ... other cases  ...
      (t "~/Downloads")))) ;; everything else
```

You can extract multiple attachments at once by prefixing the extracting command by `C-u`; so *C-u e* asks you for a range of attachments to extract (for example, *1 3-6 8*). The range "'a'" is a shortcut for *all* attachments.

5.4 Viewing images inline

It is possible to show images inline in the message view buffer if you run `emacs` in GUI-mode. You can enable this by setting the variable `mu4e-view-show-images` to `t`. Since `emacs` does not always handle images correctly, this is not enabled by default. If you are using `emacs` 24 with *ImageMagick*[3] support, make sure you call `imagemagick-register-types` in your configuration, so it is used for images.

```
;; enable inline images
(setq mu4e-view-show-images t)
;; use imagemagick, if available
(when (fboundp 'imagemagick-register-types)
    (imagemagick-register-types))
```

5.5 Displaying rich-text messages

`mu4e` normally prefers the plain-text version for messages that consist of both a plain-text and html (rich-text) versions of the body-text. You can change this by setting `mu4e-view-prefer-html` to `t`.

If there is only an html-version, or if the plain-text version is too short in comparison with the html part[4], `mu4e` tries to convert the html into plain-text for display. The default way to do that is to use the `emacs` built-in `html2text` function. However, you can set the variable `mu4e-html2text-command` to some external program instead. This program is expected to take html from standard input and write plain text in UTF-8 encoding on standard output.

An example of such a program is the program that is actually *called* `html2text`[5]. After installation, you can set it up with something like the following:

---

[2] sadly, often `application/octet-stream` is used for the mime-type, even if a better type is available

[3] http://www.imagemagick.org

[4] this is for the case where the text-part only warns that you should use the html-version

[5] http://www.mbayer.de/html2text/

```
(setq mu4e-html2text-command "html2text -utf8 -width 72")
```

An alternative to this is the Python `python-html2text` package; after installing that, you can tell `mu4e` to use it with something like:

```
(setq mu4e-html2text-command
    "html2markdown | grep -v ' _place_holder;'")
```

On MacOS, there is a program called `textutil` as yet another alternative:

```
(setq mu4e-html2text-command
    "textutil -stdin -format html -convert txt -stdout")
```

5.6 Crypto

The `mu4e` message view supports[6] decryption of encrypted messages, as well as verification of signatures. For signing/encrypting messages your outgoing messages, see Section 6.5 [Signing and encrypting], page 25.

Currently, only PGP/MIME is supported; PGP-inline and S/MIME are not.

For all of this to work, `gpg-agent` must be running, and it must set the environment variable `GPG_AGENT_INFO`. You can check from `emacs` with `M-x getenv GPG_AGENT_INFO`.

In many mainstream Linux/Unix desktop environments, everything works out-of-the-box, but if your environment does not automatically start `gpg-agent`, you can do so by hand:

```
$ eval $(gpg-agent --daemon)
```

This starts the daemon, and sets the environment variable.

5.6.1 Decryption

If you receive messages that are encrypted (using PGP/MIME), `mu4e` can try to decrypt them, base on the setting of `mu4e-decryption-policy`. If you set it to `t`, `mu4e` attempts to decrypt messages automatically; this is the default. If you set it to `nil`, `mu4e` *won't* attempt to decrypt anything. Finally, if you set it to `'ask`, it asks you what to do, each time an encrypted message is encountered.

When opening an encrypted message, `mu` consults `gpg-agent` to see if it already has unlocked the key needed to decrypt the message; if not, it prompts you for a password (typically with a separate top-level window). This is only needed once per session.

5.6.2 Verifying signatures

Some e-mail messages are cryptographically signed, and `mu4e` can check the validity of these signatures. If a message has one or more signatures, the message view shows an extra header **Signature:** (assuming it is part of your `mu4e-view-fields`), and one or more 'verdicts' of the signatures found; either **verified**, **unverified** or **error**. For instance:

Signature: unverified (Details)

You can see the details of the signature verification by activating the **Details** or pressing v. This pops up a little window with the details of the signatures found and whether they could be verified or not.

For more information, see the `mu-verify` manual page.

---

[6] Crypto-support in `mu4e` requires `mu` to have been build with crypto-support; see the Appendix C [FAQ], page 51

5.7 Custom headers

Sometimes the normal headers that mu4e offers (Date, From, To, Subject etc.) may not be enough. For these cases, mu4e offers *custom headers* in both the headers-view and the message-view.

See Section 4.5 [HV Custom headers], page 15 for an example of this; the difference for the message-view is that you should add your custom header to mu4e-view-fields rather than mu4e-headers-fields.

5.8 Actions

You can perform custom functions ("actions") on messages and their attachments. For a general discussion on how to define your own, see see Chapter 10 [Actions], page 37.

5.8.1 Message actions

mu4e-view-action (a) lets you pick some custom action to perform on the current message. You can specify these actions using the variable mu4e-view-actions; mu4e defines a number of example actions.

5.8.2 Attachment actions

Similarly, there is mu4e-view-attachment-action (A) for actions on attachments, which you can specify with mu4e-view-attachment-actions.

mu4e predefines a number of attachment-actions:

- open-with (w): open the attachment with some arbitrary program. For example, suppose you have received a message with a picture attachment; then, *A w 1 RET gimp RET* opens that attachment in *The Gimp*

- pipe (|: process the attachment with some Unix shell-pipe and see the results. Suppose you receive a patch file, and would like to get an overview of the changes, using the diffstat program. You can use something like: *A | 1 RET diffstat -b RET*.

- emacs (e): open the attachment in your running emacs. For example, if you receive some text file you'd like to open in emacs: *A e 1 RET*.

These actions all work on a *temporary copy* of the attachment.

6 The editor view

Writing e-mail messages takes place in the Editor View. mu4e's editor view builds on top of Gnu's `message-mode`. Most of the `message-mode` functionality is available, as well some mu4e-specifics. Its major mode is `mu4e-compose-mode`.

6.1 Overview

```
From: Rupert the Monkey <rupert@example.com>
To: Wally the Walrus <wally@example.com>
Subject: Re: Eau-qui d'eau qui?
--text follows this line--

On Mon 16 Jan 2012 10:18:47 AM EET, Wally the Walrus wrote:

 > Hi Rupert,
 >
 > Dude - how are things?
 >
 > Later -- wally.
```

6.2 Useful keybindings

mu4e's editor view derives from Gnu's message editor and shares most of its keybindings. Here are some of the more useful ones (you can use the menu to find more):

```
key            description
---            -----------
C-c C-c        send message
C-c C-d        save to drafts and leave
C-c C-k        kill the message
C-c C-a        attach a file (pro-tip: drag & drop works as well)

(mu4e-specific)
C-S-u          update mail & reindex
```

6.3 Address autocompletion

mu4e supports[1] autocompleting addresses when composing e-mail messages. mu4e uses the e-mail addresses from the messages you sent or received as the source for this. Address auto-completion is enabled by default; if you want to disable it for some reason, set `mu4e-compose-complete-addresses` to `nil`.

Emacs 24 also supports cycling through the alternatives. When there are more than 5 matching addresses, they are shown in a `*Completions*` buffer. Once the number of matches gets below this number, one is inserted in the address field and you can cycle through the alternatives using `TAB`.

---

[1] `emacs` 23.2 or higher is required

6.3.1 Limiting the number of addresses

If you have a lot of mail, especially from mailing lists and the like, there can be a *lot* of e-mail addresses, many of which may not be very useful when auto-completing. For this reason, mu4e attempts to limit the number of e-mail addresses in the completion pool by filtering out the ones that are not likely to be relevant. The following variables are available for tuning this:

- `mu4e-compose-complete-only-personal` - when set to `t`, only consider addresses that were seen in *personal* messages – that is, messages in which one of my e-mail addresses was seen in one of the address fields. This is to exclude mailing list posts. You can define what is considered 'my e-mail address' using `mu4e-user-mail-address-list`, a list of e-mail address (defaults to `user-mail-address`, and when indexing from the command line, the `--my-address` parameter for `mu index`.

- `mu4e-compose-complete-only-after` - only consider e-mail addresses last seen after some date. Parameter is a string, parseable by `org-parse-time-string`. This excludes old e-mail addresses. The default is `"2010-01-01"`, i.e., only consider e-mail addresses seen since the start of 2010.

- `mu4e-compose-complete-ignore-address-regexp` - a regular expression to filter out other 'junk' e-mail addresses; defaults to "`no-?reply`".

6.4 Compose hooks

If you want to change some setting, or execute some custom action before message composition starts, you can define a *hook function*. mu4e offers two hooks:

- `mu4e-compose-pre-hook`: this hook is run *before* composition starts; if you are composing a *reply*, *forward* a message, or *edit* an existing message, the variable `mu4e-compose-parent-message` points to the message being replied to, forwarded or edited, and you can use `mu4e-message-field` to get the value of various properties (and see Section 11.3 [Message functions], page 40).

- `mu4e-compose-mode-hook`: this hook is run just before composition starts, when the whole buffer has already been set up. This is a good place for editing-related settings. `mu4e-compose-parent-message` (see above) is also at your disposal.

Let's look at some examples. First, suppose we want to set the `From:`-address for a reply message based on the receiver of the original:

```
;; 1) messages to me@foo.com should be replied with From:me@foo.com
;; 2) messages to me@bar.com should be replied with From:me@bar.com
;; 3) all other mail should use From:me@cuux.com
(add-hook 'mu4e-compose-pre-hook
  (defun my-set-from-address ()
    "Set the From address based on the To address of the original."
    (let ((msg mu4e-compose-parent-message)) ;; msg is shorter...
      (when msg
(setq user-mail-address
  (cond
    ((mu4e-message-contact-field-matches msg :to "me@foo.com")
      "me@foo.com")
```

```
        ((mu4e-message-contact-field-matches msg :to "me@bar.com")
          "me@bar.com")
        (t "me@cuux.com")))))))
```

Second, as mentioned, `mu4e-compose-mode-hook` is especially useful for editing-related settings. For example:

```
(add-hook 'mu4e-compose-mode-hook
  (defun my-do-compose-stuff ()
    "My settings for message composition."
    (set-fill-column 72)
    (flyspell-mode)))
```

This hook is also useful for adding headers or changing headers, since the message is fully formed when this hook runs. For example, to add a `Bcc:`-header, you could add something like the following, using `message-add-header` from `message-mode`.

```
(add-hook 'mu4e-compose-mode-hook
  (defun my-add-bcc ()
    "Add a Bcc: header."
    (save-excursion (message-add-header "Bcc: me@example.com\n"))))
```

For a more general discussion about extending `mu4e`, see Chapter 11 [Extending mu4e], page 39.

6.5 Signing and encrypting

Signing and encrypting of messages is possible using `emacs-mime` (See Info file `emacs-mime`, node 'Composing'), most easily accessed through the `Attachments`-menu while composing a message, or with *M-x mml-secure-message-encrypt-pgp*, *M-x mml-secure-message-sign-pgp*.

The support for encryption and signing is *independent* of the support for their counterparts, decrypting and signature verification (as discussed in Section 5.6 [MSGV Crypto], page 21). Even if your `mu4e` does not have support for the latter two, you can still sign/encrypt messages.

Currently, decryption and signature verification only works for PGP/MIME; inline-PGP and S/MIME are not supported.

6.6 Queuing mail

If you cannot send mail right now, for example because you are currently offline, you can *queue* the mail, and send it when you have restored your internet connection. You can control this from the Chapter 3 [Main view], page 10.

To allow for queuing, you need to tell `smtpmail` where you want to store the queued messages. For example:

```
(setq smtpmail-queue-mail t  ;; start in queuing mode
      smtpmail-queue-dir   "~/Maildir/queue/cur")
```

For convenience, we put the queue directory somewhere in our normal maildir. If you want to use queued mail, you should create this directory before starting `mu4e`. The `mu mkdir` command may be useful here, so for example:

```
$ mu mkdir ~/Maildir/queue
$ touch ~/Maildir/queue/.noindex
```

The file created by the `touch` command tells `mu` to ignore this directory for indexing, which makes sense since it contains `smtpmail` meta-data rather than normal messages; see the `mu-mkdir` and `mu-index` man-pages for details.

Warning: when you switch on queued-mode, your messages *won't* reach their destination until you switch it off again; so, be careful not to do this accidentally!

6.7 Message signatures

Message signatures are the standard footer blobs in e-mail messages where you can put in information you want to include in every message. The text to include is set with `mu4e-compose-signature`.

If you don't want to include this automatically with each message, you can set `mu4e-compose-signature-auto-include` to `nil`; you can then still include the signature manually, using the function `message-insert-signature`, typically bound to *C-c C-w*.

6.8 Other settings

- If you want use `mu4e` as `emacs`' default program for sending mail, see Section A.1 [Setting the default emacs mail program], page 41.
- Normally, `mu4e` *buries* the message buffer after sending; if you want to kill the buffer instead, add something like the following to your configuration:

  ```
  (setq message-kill-buffer-on-exit t)
  ```

7 Searching

mu4e is fully search-based: even if you 'jump to a folder', you are executing a query for messages that happen to have the property of being in a certain folder (maildir).

Normally, queries return up to `mu4e-headers-results-limit` (default: 500) results. That is usually more than enough, and makes things significantly faster. Sometimes, however, you may want to show *all* results; you can enable this with *M-x mu4e-headers-toggle-full-search*, or by customizing the variable `mu4e-headers-full-search`. This applies to all search commands.

You can also influence the sort order and whether threads are shown or not; see Section 4.4 [Sort order and threading], page 15.

7.1 Queries

mu4e queries are the same as the ones that `mu find` understands[1]. Let's look at some examples here, please refer to the `mu-find` and `mu-easy` man pages for details and more examples.

- Get all messages regarding *bananas*:

  `bananas`
- Get all messages regarding *bananas* from *John* with an attachment:

  `from:john flag:attach bananas`
- Get all messages with subject *wombat* in June 2009

  `subject:wombat date:20090601..20090630`
- Get all messages with PDF attachments in the `/projects` folder

  `maildir:/projects mime:application/pdf`
- Get all messages about *Rupert* in the `/Sent Items` folder. Note that maildirs with spaces must be quoted.

  `maildir:"/Sent Items" rupert`
- Get all important messages which are signed:

  `flag:signed prio:high`
- Get all messages from *Jim* without an attachment:

  `from:jim AND NOT flag:attach`
- Get all messages with Alice in one of the contacts-fields (`to`, `from`, `cc`, `bcc`):

  `contact:alice`
- Get all unread messages where the subject mentions ngstrm: (search is case-insensitive and accent-insensitive, so this matches ngstrm, angstrom, aNGstrM, ...)

  `subject:ngstrm flag:unread`
- Get all unread messages between Mar-2002 and Aug-2003 about some bird:

  `date:20020301..20030831 nightingale flag:unread`
- Get today's messages:

---

[1] with the caveat that command-line queries are subject to the shell's interpretation before `mu` sees them

```
date:today..now
```

or, unless you have a really old Xapian

```
date:today
```

- Get all messages we got in the last two weeks regarding *emacs*:

```
date:2w..now emacs
```

or, unless you have a really old Xapian

```
date:2w.. emacs
```

- Get messages from the *Mu* mailing list:

```
list:mu-discuss.googlegroups.com
```

- Get messages with a subject soccer, Socrates, society, ...; note that the '*'-wildcard can only appear as a term's rightmost character:

```
subject:soc*
```

- Get all messages *not* sent to a mailing-list:

```
NOT flag:list
```

- Get all mails with attachments with filenames starting with *pic*; note that the '*' wildcard can only appear as the term's rightmost character:

```
file:pic*
```

- Get all messages with PDF-attachments:

```
mime:application/pdf
```

Get all messages with image attachments, and note that the '*' wildcard can only appear as the term's rightmost character:

```
mime:image/*
```

7.2 Bookmarks

If you have queries that you use often, you may want to store them as *bookmarks*. Bookmark searches are available in the main view Chapter 3 [Main view], page 10, header view See Chapter 4 [Headers view], page 12, and message view See Chapter 5 [Message view], page 17, using (by default) the key b (*M-x mu4e-search-bookmark*), or B (*M-x mu4e-search-bookmark-edit*) which lets you edit the bookmark first.

7.2.1 Setting up bookmarks

mu4e provides a number of default bookmarks. Their definition is instructive:

```
(defvar mu4e-bookmarks
  '( ("flag:unread AND NOT flag:trashed" "Unread messages"      ?u)
     ("date:today..now"                  "Today's messages"     ?t)
     ("date:7d..now"                     "Last 7 days"          ?w)
     ("mime:image/*"                     "Messages with images" ?p))
  "A list of pre-defined queries; these show up in the main
screen. Each of the list elements is a three-element list of the
form (QUERY DESCRIPTION KEY), where QUERY is a string with a mu
query, DESCRIPTION is a short description of the query (this
shows up in the UI), and KEY is a shortcut key for the query.")
```

You can replace these or add your own items, by putting in your configuration (`~/.emacs`) something like:

```
(add-to-list 'mu4e-bookmarks
  '("size:5M..500M"        "Big messages"      ?b))
```

This prepends your bookmark to the list, and assigns the key b to it. If you want to *append* your bookmark, you can use t as the third argument to `add-to-list`.

In the various mu4e views, pressing b lists all the bookmarks defined in the echo area, with the shortcut key highlighted. So, to invoke the bookmark we just defined (to get the list of "Big Messages"), all you need to type is *bb*.

7.2.2 Editing bookmarks before searching

There is also *M-x mu4e-headers-search-bookmark-edit* (key B), which lets you edit the bookmarked query before invoking it. This can be useful if you have many similar queries, but need to change some parameter. For example, you could have a bookmark '`"date:today..now AND "`'[2], which limits any result to today's messages.

7.3 Maildir searches

Maildir searches are quite similar to bookmark searches (see Section 7.2 [Bookmarks], page 28), with the difference being that the target is always a maildir – maildir queries provide a 'traditional' folder-like interface to a search-based e-mail client. By default, maildir searches are available in the Chapter 3 [Main view], page 10, Chapter 4 [Headers view], page 12, and Chapter 5 [Message view], page 17, with the key j (`mu4e-jump-to-maildir`).

7.3.1 Setting up maildir shortcuts

You can search for maildirs like can for any other messsage property (e.g. with a query like `maildir:/myfolder`), but since it is so common, mu4e offers a shortcut for this.

For this to work, you need to set the variable `mu4e-maildir-shortcuts` to the list of maildirs you want to have quick access to, for example:

```
(setq mu4e-maildir-shortcuts
 '( ("/inbox"       . ?i)
    ("/archive"     . ?a)
    ("/lists"       . ?l)
    ("/work"        . ?w)
    ("/sent"        . ?s)))
```

This sets i as a shortcut for the /inbox folder – effectively a query `maildir:/inbox`. There is a special shortcut o or / for *other* (so don't use those for your own shortcuts!), which allows you to choose from *all* maildirs that you have. There is support for autocompletion; note that the list of maildirs is determined when mu4e starts; if there are changes in the maildirs while mu4e is running, you need to restart mu4e.

Each of the folder names is relative to your top-level maildir directory; so if you keep your mail in ~/Maildir, /inbox would refer to ~/Maildir/inbox. With these shortcuts, you can jump around your maildirs (folders) very quickly - for example, getting to the /lists folder only requires you to type *jl*, then change to /work with *jw*.

---

[2] Not a valid search query by itself

While in queries you need to quote folder names (maildirs) with spaces in them, you should *not* quote them when used in `mu4e-maildir-shortcuts`, since `mu4e` does that automatically for you.

The very same shortcuts are used by `M-x mu4e-mark-for-move` (default shortcut m); so, for example, if you want to move a message the `/archive` folder, you can do so by typing `ma`.

7.4 Other search functionality

7.4.1 Navigating through search queries

You can navigate through previous/next queries using `mu4e-headers-query-prev` and `mu4e-headers-query-next`, which are bound to `M-left` and `M-right`, similar to what some web browsers do.

`mu4e` tries to be smart and not record duplicate queries. Also, the number of queries remembered has a fixed limit, so `mu4e` won't use too much memory, even if used for a long time. However, if you want to forget previous/next queries, you can use `M-x mu4e-headers-forget-queries`.

7.4.2 Narrowing search results

It can be useful to narrow existing search results, that is, to add some clauses to the current query to match fewer messages.

For example, suppose you're looking at some mailing list, perhaps by jumping to a maildir (`M-x mu4e-headers-jump-to-maildir`, j) or because you followed some bookmark (`M-x mu4e-headers-search-bookmark`, b). Now, you want to narrow things down to only those messages that have attachments.

This is when `M-x mu4e-headers-search-narrow` (/) comes in handy. It asks for an additional search pattern, which is appended to the current search query, in effect getting you the subset of the currently shown headers that also match this extra search pattern. \ takes you back to the previous query, so, effectively 'widens' the search. Technically, narrowing the results of query x with expression y implies doing a search `(x) AND y`.

Note, messages that were not in your in your original search results because of `mu4e-headers-results-limit`, may show up in the narrowed query.

7.4.3 Including related messages

It can be useful to not only show the messages that directly match a certain query, but also include messages that are related to these messages. That is, messages that belong to the same discussion threads are included in the results, just like e.g. Gmail does it. You can enable this behavior by setting `mu4e-headers-include-related` to t, and you can toggle between including/not-including with W.

Be careful though when e.g. deleting ranges of messages from a certain folder – the list may now also include messages from *other* folders.

7.4.4 Skipping duplicates

Another useful feature is skipping of *duplicate messages*. When you have copies of messages, there's usually little value in including more than one in search results. A common reason for having multiple copies of messages is the combination of Gmail and `offlineimap`, since that is the way the labels / virtual folders in Gmail are represented. You can enable skipping duplicates by setting `mu4e-headers-skip-duplicates` to `t`, and you can toggle between the skipping/not skipping with `V`.

Note, messages are considered duplicates when they have the same `Message-Id`.

8 Marking

In mu4e, the common way to do things with messages is a two-step process - first you *mark* them for a certain action, then you *execute* (**x**) those marks. This is similar to the way dired operates. Marking can happen in both the Chapter 4 [Headers view], page 12 and the Chapter 5 [Message view], page 17.

8.1 Selecting messages for marking

There are multiple ways to mark messages:

- *message at point*: you can put a mark on the message-at-point in either the Chapter 4 [Headers view], page 12 or Chapter 5 [Message view], page 17

- *region*: you can put a mark on all messages in the current region (selection) in the Chapter 4 [Headers view], page 12

- *pattern*: you can put a mark on all messages in the Chapter 4 [Headers view], page 12 matching a certain pattern with *M-x mu4e-headers-mark-pattern* (%)

- *thread/subthread*: You can put a mark on all the messages in the thread/subthread at point with *M-x mu4e-headers-mark-thread* and *M-x mu4e-headers-mark-subthread*, respectively

8.2 What to mark for

mu4e supports a number of marks:

```
mark for/as  | keybinding  | description
-------------+-------------+-------------------------
'something'  | <insert>    | mark now, decide later
delete       | D, <delete> | delete
flag         | +           | mark as 'flagged' ('starred')
move         | m           | move to some maildir
read         | !           | mark as read
refile       | r           | mark for refiling
trash        | d           | move to the trash folder
untrash      | =           | remove 'trash' flag
unflag       | -           | remove 'flagged' mark
unmark       | u           | remove mark at point
unmark all   | U           | remove all marks
unread       | ?           | marks as unread
```

After marking a message, the left-most columns in the headers view indicate the kind of mark. This is informative, but if you mark many (say, thousands) messages, this slows things down significantly[1]. For this reason, you can disable this by setting mu4e-headers-show-target to nil.

---

[1] this uses an emacs feature called *overlays*, which are slow when used a lot in a buffer

`something` is a special kind of mark; you can use it to mark messages for 'something', and then decide later what the 'something' should be[2] Later, you can set the actual mark using *M-x mu4e-mark-resolve-deferred-marks* (#). Alternatively, `mu4e` will ask you when you try to execute the marks (x).

8.3 Executing the marks

After you have marked some messages, you can execute them with x (*M-x mu4e-mark-execute-all*).

8.4 Leaving the headers buffer

When you quit or update a headers buffer that has marked messages (for example, by doing a new search), `mu4e` asks you what to do with them, depending on the value of the variable `mu4e-headers-leave-behavior` – see its documentation.

8.5 Built-in marking functions

Some examples of `mu4e`'s built-in marking functions.

- *Mark the message at point for trashing*: press d
- *Mark all messages in the buffer as unread*: press *C-x h o*
- *Delete the messages in the current thread*: press *T D*
- *Mark messages with a subject matching "hello" for flagging*: press *% s hello RET*.

8.6 Custom mark functions

Sometimes, the built-in functions to mark messages may not be sufficient for your needs. For this, `mu4e` offers an easy way to define your own custom mark functions. You can choose one of the custom marker functions by pressing & in the Chapter 4 [Headers view], page 12 and Chapter 5 [Message view], page 17.

Custom mark functions are to be appended to the list `mu4e-headers-custom-markers`. Each of the elements of this list ('markers') is a list with two or three elements:

1. The name of the marker - a short string describing this marker. The first character of this string determines its shortcut, so these should be unique. If necessary, simply prefix the name with a unique character.

2. a predicate function, taking two arguments *msg* and *param*. *msg* is the message plist (see Section 11.3 [Message functions], page 40 and *param* is a parameter provided by the third of the marker elements (see the next item). The predicate function should return non-`nil` if the message matches.

3. (optionally) a function that is evaluated once, and the result is passed as a parameter to the predicate function. This is useful when user-input is needed.

Let's look at an example: suppose we want to match all messages that have more than *n* recipients – we could do this with the following recipe:

---

[2] This kind of 'deferred marking' is similar to the facility in **midnight commander** (http://www.midnight-commander.org/) and the like, and uses the same key binding (**insert**).

```
(add-to-list 'mu4e-headers-custom-markers
  '("More than n recipients"
      (lambda (msg n)
        (> (+ (length (mu4e-message-field msg :to))
              (length (mu4e-message-field msg :cc))) n))
      (lambda ()
        (read-number "Match messages with more recipients than: "))) t)
```

After evaluating this expression, you can use it by pressing & in the headers buffer to select a custom marker function, and then M to choose this particular one (M because it is the first character of the description).

As you can see, it's not very hard to define simple functions to match messages. There are more examples in the defaults for **mu4e-headers-custom-markers**; see **mu4e-headers.el** and see Chapter 11 [Extending mu4e], page 39 for general information about writing your own functions.

9 Dynamic folders

In Section 2.6 [Folders], page 7, we explained how you can set up mu4e's special folders:

```
(setq
  mu4e-sent-folder   "/sent"     ;; sent messages
  mu4e-drafts-folder "/drafts"   ;; unfinished messages
  mu4e-trash-folder  "/trash"    ;; trashed messages
  mu4e-refile-folder "/archive") ;; saved messages
```

In some cases, having such static folders may not suffice - perhaps you want to change the folders depending on the context. For example, the folder for refiling could vary, based on the sender of the message.

To make this possible, instead of setting the standard folders to a string, you can set them to be a *function* that takes a message as its parameter, and returns the desired folder name. This chapter shows you how to do that. For a more general discussion of how to extend mu4e and writing your own functions, see Chapter 11 [Extending mu4e], page 39.

9.1 Smart refiling

When refiling messages, perhaps to archive them, it can be useful to have different target folders for different messages, based on some property of those message – smart refiling.

To accomplish this, we can set the refiling folder (`mu4e-refile-folder`) to a function that returns the actual refiling folder for the particular message. An example should clarify this:

```
(setq mu4e-refile-folder
  (lambda (msg)
    (cond
      ;; messages to the mu mailing list go to the /mu folder
      ((mu4e-message-contact-field-matches msg :to
        "mu-discuss@googlegroups.com")
       "/mu")
      ;; messages sent directly to me go to /archive
      ;; also `mu4e-user-mail-address-p' can be used
      ((mu4e-message-contact-field-matches msg :to "me@example.com")
        "/private")
      ;; messages with football or soccer in the subject go to /football
      ((string-match "football\\|soccer"
        (mu4e-message-field msg :subject))
        "/football")
      ;; messages sent by me go to the sent folder
      ((find-if
  (lambda (addr)
    (mu4e-message-contact-field-matches msg :from addr))
  mu4e-user-mail-address-list)
 mu4e-sent-folder)
      ;; everything else goes to /archive
      ;; important to have a catch-all at the end!
```

```
(t  "/archive"))))
```

This can be very powerful; you can select some messages in the headers view, then press r, and have them all marked for refiling to their particular folders.

Some notes:

- We set **mu4e-refile-folder** to an anonymous (**lambda**) function. This function takes one argument, a message plist[1]. The plist corresponds to the message at point. See Section 11.3 [Message functions], page 40 for a discussion on how to deal with them.

- In our function, we use a **cond** control structure; the function returns the first of the clauses that matches. It's important to make the last clause a catch-all, so we always return *some* folder.

- We use the convenience function **mu4e-message-contact-field-matches**, which evaluates to **t** if any of the names or e-mail addresses in a contact field (in this case, the **To:**-field) matches the regular expression.

9.2 Other dynamic folders

Using the same mechanism, you can create dynamic sent-, trash-, and drafts-folders. The message-parameter you receive for the sent and drafts folder is the *original* message, that is, the message you reply to, or forward, or edit. If there is no such message (for example when composing a brand new message) the message parameter is **nil**.

Let's look at an example. Suppose you want a different trash folder for work-email. You can achieve this with something like:

```
(setq mu4e-trash-folder
    (lambda (msg)
        ;; the 'and msg' is to handle the case where msg is nil
        (if (and msg
                (mu4e-message-contact-field-matches msg :to "me@work.com"))
        "/trash-work"
        "/trash")))
```

Good to remember:

- The *msg* parameter you receive in the function refers to the *original message*, that is, the message being replied to or forwarded. When re-editing a message, it refers to the message being edited. When you compose a totally new message, the *msg* parameter is **nil**.

- When re-editing messages, the value of **mu4e-drafts-folder** is ignored.

---

[1] a property list describing a message

10 Actions

mu4e lets you define custom actions for messages in the Chapter 4 [Headers view], page 12 and for both messages and attachments in the Chapter 5 [Message view], page 17. Custom actions allow you to easily extend mu4e for specific needs – for example, marking messages as spam in a spam filter or applying an attachment with a source code patch.

You can invoke the actions with key a for actions on messages, and key A for actions on attachments.

For general information extending mu4e and writing your own functions, see Chapter 11 [Extending mu4e], page 39.

10.1 Defining actions

Defining a new custom action comes down to writing an elisp-function to do the work. Functions that operate on messages receive a *msg* parameter, which corresponds to the message at point. Something like:

```
(defun my-action-func (msg)
 "Describe my message function."
;; do stuff
)
```

Messages that operate on attachments receive a *msg* parameter, which corresponds to the message at point, and an *attachment-num*, which is the number of the attachment as seen in the message view. An attachment function looks like:

```
(defun my-attachment-action-func (msg attachment-num)
 "Describe my attachment function."
;; do stuff
)
```

After you have defined your function, you can add it to the list of actions[1], either mu4e-headers-actions, mu4e-view-actions or mu4e-view-attachment-actions. The format[2] of each action is a cons-cell, (DESCRIPTION . VALUE); see below for some examples. If your shortcut is not also the first character of the description, simply prefix the description with that character.

Let's look at some examples.

10.2 Adding an action in the headers view

Suppose we want to inspect the number of recipients for a message in the Chapter 4 [Headers view], page 12. We add the following to our configuration:

```
(defun show-number-of-recipients (msg)
  "Display the number of recipients for the message at point."
  (message "Number of recipients: %d"
```

---

[1] Instead of defining the functions separately, you can obviously also add a lambda-function directly to the list; however, separate functions are easier to change

[2] Note, the format of the actions has changed since version 0.9.8.4, and you must change your configuration to use the new format; mu4e warns you when you are using the old format.

```
    (+ (length (mu4e-message-field msg :to))
       (length (mu4e-message-field msg :cc)))))))


;; define 'N' (the first letter of the description) as the shortcut
;; the 't' argument to add-to-list puts it at the end of the list
(add-to-list 'mu4e-headers-actions
    '("Number of recipients" . show-number-of-recipients) t)
```

After evaluating this, a *N* in the headers view shows the number of recipients for the message at point.

10.3 Adding an action in the message view

As another example, suppose we would like to search for messages by the sender of the message at point:

```
(defun search-for-sender (msg)
   "Search for messages sent by the sender of the message at point."
   (mu4e-headers-search
      (concat "from:" (cdar (mu4e-message-field msg :from)))))


;; define 'x' as the shortcut
(add-to-list 'mu4e-view-actions
    '("xsearch for sender" . search-for-sender) t)
```

If you wonder why we use `cdar`, remember that the `From:`-field is a list of (`NAME` . `EMAIL`) cells; thus, `cdar` gets us the e-mail address of the first in the list. `From:`-fields rarely contain multiple cells.

10.4 Adding an attachment action

Finally, let's define an attachment action. As mentioned, attachment-action functions receive *2* arguments, the message and the attachment number to use.

The following example action counts the number of lines in an attachment, and defines n as its shortcut key (the n is prefixed to the description).

```
(defun count-lines-in-attachment (msg attachnum)
   "Count the number of lines in an attachment."
   (mu4e-view-pipe-attachment msg attachnum "wc -l"))


;; defining 'n' as the shortcut
(add-to-list 'mu4e-view-attachment-actions
    '("ncount lines" . count-lines-in-attachment) t)
```

10.5 More example actions

mu4e includes a number of example actions in the file `mu4e-actions.el` in the source distribution (see *C-h f mu4e-action-TAB*). For example, for viewing messages in an external web browser, or listening to a message's body-text using text-to-speech.

11 Extending mu4e

mu4e is designed to be easily extendible - that is, write your own emacs-lisp to make mu4e behave exactly as you want. Here, we provide some guidelines for doing so.

11.1 Extension points

There are a number of places where mu4e lets you plug in your own functions:

- Custom functions for for message headers in the message-view and headers-view - see Section 4.5 [HV Custom headers], page 15, Section 5.7 [MSGV Custom headers], page 22

- Using message-specific folders for drafts, trash, sent messages and refiling, based on a function - see Chapter 9 [Dynamic folders], page 35

- Using an attachment-specific download-directory - see the variable mu4e-attachment-dir.

- Apply a function to a message in the headers view - see Section 10.2 [Adding an action in the headers view], page 37

- Apply a function to a message in the message view - see Section 10.3 [Adding an action in the message view], page 38

- Apply a function to to an attachment - see Section 10.4 [Adding an attachment action], page 38

- Custom function to mark certain messages - see Section 8.6 [Custom mark functions], page 33

- Using various *mode*-hooks, mu4e-compose-pre-hook (see Section 6.4 [Compose hooks], page 24), mu4e-index-updated-hook (see Appendix C [FAQ], page 51)

You can also write your own functions without using the above. If you want to do so, key useful functions are mu4e-message-at-point (see below), mu4e-headers-for-each (to iterate over all headers, see its docstring) and mu4e-view-for-each-part (to iterate over all parts/attachments, see its docstring).

11.2 Available functions

The whole of mu4e consists of hundreds of elisp functions. However, the majority of those are for *internal* use only; you can recognize them easily, because they all start with mu4e~. These function make all kinds of assumptions, and they are subject to change, and should therefore *not* be used. The same is true for *variables* that start with mu4e~; don't touch them. Let me repeat that:

    Do not use mu4e~... functions or variables!

In addition, you should use functions in the right context; functions that start with mu4e-view- are only applicable to the message view, while functions starting with mu4e-headers- are only applicable to the headers view. Functions without such prefixes are applicable everywhere.

11.3 Message functions

Many functions in `mu4e` deal with message plist (property lists). They contain information about messages, such as sender and recipient, subject, date and so on. To deal with these plists, there are a number of `mu4e-message-` functions (in `mu4e-message.el`), such as `mu4e-message-field` and `mu4e-message-at-point`

For example, to get the subject of the message at point, in either the headers view or the message view, you could write:

```
(mu4e-message-field (mu4e-message-at-point) :subject)
```

Note that:

- The contact fields (To, From, Cc, Bcc) are lists of cons-pairs (`name . email`); `name` may be `nil`. So, for example:
  ```
  (mu4e-message-field some-msg :to)
  ;; => (("Jack" . "jack@example.com") (nil . "foo@example.com"))
  ```
  If you are only looking for a match in this list (e.g., "Is Jack one of the recipients of the message?"), there is a convenience function `mu4e-message-contact-field-matches` to make this easy.

- The message body is only available in the message view, not in the headers view.

11.4 Utility functions

`mu4e-utils` contains a number of utility functions; we list a few here; see their docstrings for the details:

- `mu4e-read-option`: read one option from a list. For example:
  ```
  (mu4e-read-option "Choose an animal: "
      '(("Monkey" . monkey) ("Gnu" . gnu) ("xMoose" . moose)))
  ```
  The user is presented with:
  ```
  Choose an animal: [M]onkey, [G]nu, [x]Moose
  ```
- `mu4e-ask-maildir`: ask for a maildir; try one of the shortcuts (`mu4e-maildir-shortcuts`), or the full set of available maildirs.
- `mu4e-running-p`: return `t` if the `mu4e` process is running, `nil` otherwise.
- (`mu4e-user-mail-address-p addr`): return `t` if *addr* is one of the user's e-mail addresses (as per `mu4e-user-mail-address-list`).
- `mu4e-log` logs to the `mu4e` debugging log if it is enabled; see `mu4e-toggle-logging`.
- `mu4e-message`, `mu4e-warning`, `mu4e-error` are the `mu4e` equivalents of the normal elisp `message`, `user-error`[1] and `error` functions.

---

[1] `user-error` only appears in `emacs` 24.2 and later; in older versions it falls back to `error`

Appendix A Interaction with other tools

In this chapter, we discuss some ways in ways in which `mu4e` can coperate with other tools.

A.1 Setting the default `emacs` mail program

`emacs` allows you to select an e-mail program as the default program it uses when you press `C-x m` (`compose-mail`), call `report-emacs-bug` and so on. If you want to use `mu4e` for this, you do so by adding the following to your configuration:

```
(setq mail-user-agent 'mu4e-user-agent)
```

At the present time, support is *experimental*.

A.2 Creating `org-mode` links

It can be useful to include links to e-mail messages or even search queries in your org-mode files. `mu4e` supports this with the `org-mu4e` module; you can set it up by adding it to your configuration:

```
(require 'org-mu4e)
```

After this, you can use the normal `org-mode` mechanisms to store links: *M-x org-store-link* stores a link to a particular message when you're in Chapter 5 [Message view], page 17, and a link to a query when you are in Chapter 4 [Headers view], page 12.

You can insert this link later with *M-x org-insert-link*. From `org-mode`, you can go to the query or message the link points to with either *M-x org-agenda-open-link* in agenda buffers, or *M-x org-open-at-point* elsewhere - both typically bound to *C-c C-o*.

A.3 Rich-text messages with `org-mode` (deprecated)

Some earlier versions of `mu4e` had support for editing e-mail messages using `org-mode`; since this never worked very well, this has now been deprecated; it might be replaced in some future version with something better.

A.3.1 Some caveats

A.4 Maintaining an address-book with org-contacts

Note, `mu4e` supports built-in address autocompletion; Section 6.3 [Address autocompletion], page 23, and that is the recommended way to do this. However, it is also possible to manage your addresses with `org-mode`, using `org-contacts`[1].

`mu4e-actions` defines a useful action (Chapter 10 [Actions], page 37) for adding a contact based on the `From:`-address in the message at point. To enable this, add to your configuration something like:

```
(setq mu4e-org-contacts-file  <full-path-to-your-org-contacts-file>)
(add-to-list 'mu4e-headers-actions
  '("org-contact-add" . mu4e-action-add-org-contact) t)
(add-to-list 'mu4e-view-actions
```

---

[1] http://julien.danjou.info/software/org-contacts.el

```
                    '("org-contact-add" . mu4e-action-add-org-contact) t)
```

After this, you should be able to add contacts using a o in the headers view and the message view, using the org-capture mechanism. Note, the shortcut character o is due to the first character of org-contact-add.

A.5 Getting new mail notifications with Sauron

The emacs-package sauron[2] (by the same author) can be used to get notifications about new mails. If you run something like the below script from your crontab (or have some other way of having it execute every n minutes), you receive notifications in the sauron-buffer when new messages arrive.

```sh
#!/bin/sh

# put the path to your Inbox folder here
CHECKDIR="/home/$LOGNAME/Maildir/Inbox"

sauron-msg () {
    DBUS_COOKIE="/home/$LOGNAME/.sauron-dbus"
    if test "x$DBUS_SESSION_BUS_ADDRESS" = "x"; then
        if test -e $DBUS_COOKIE; then
                export DBUS_SESSION_BUS_ADDRESS="`cat $DBUS_COOKIE`"
        fi
    fi
    if test -n "x$DBUS_SESSION_BUS_ADDRESS"; then
        dbus-send --session                                     \
            --dest="org.gnu.Emacs"                              \
            --type=method_call                                  \
            "/org/gnu/Emacs/Sauron"                             \
            "org.gnu.Emacs.Sauron.AddMsgEvent"                  \
            string:shell uint32:3 string:"$1"
    fi
}

#
# -mmin -5: consider only messages that were created / changed in the
# the last 5 minutes
#
for f in `find $CHECKDIR -mmin -5 -a -type f`; do
        subject=`$MU view $f | grep '^Subject:' | sed 's/^Subject://'`
        sauron-msg "mail: $subject"
done
```

You might want to put:

```
    (setq sauron-dbus-cookie t)
```

---

[2] Sauron can be found at https://github.com/djcb/sauron, or in the Marmalade package-repository at http://http://marmalade-repo.org/

in your setup, to allow the script to find the D-Bus session bus, even when running outside its session.

A.6 Speedbar support

`speedbar` is an `emacs`-extension that shows navigational information for an `emacs` buffer in a separate frame. Using `mu4e-speedbar`, `mu4e` lists your bookmarks and maildir folders and allows for one-click access to them.

`mu4e` loads `mu4e-speedbar` automatically; all you need to do to activate it is *M-x speedbar*. Then, when then switching to the Chapter 3 [Main view], page 10, the speedbar-frame is updated with your bookmarks and maildirs. For speed reasons, the list of maildirs is determined when `mu4e` starts; if the list of maildirs changes while `mu4e` is running, you need to restart `mu4e` to have those changes reflected in the speedbar and in other places that use this list, such as auto-completion when jumping to a maildir.

`mu4e-speedbar` was contributed by *Antono Vasiljev*.

A.7 Citations with `mu-cite`

`mu-cite`[3] is a package to control the way message citations look like (i.e., the message you responded to when you reply to them or forward them), with its latest version available at `http://www.jpl.org/elips/mu/`.

After installing `mu-cite`, you can use something like the following to make it work with `mu4e`:

```
(require 'mu-cite)
(setq mu4e-cite-function 'mu-cite-original)
(setq mu-cite-top-format
    '("On " date ", " from " wrote:\n\n"))
(setq mu-cite-prefix-format '(" > ")))
```

A.8 Attaching files with `dired`

It is possible to attach files to `mu4e` messages using `dired` (See Info file `emacs`, node 'Dired'), using the following steps (based on a post on the `mu-discuss` mailing list by *Stephen Eglen*).

To prepare for this, you need a special version of the `gnus-dired-mail-buffers` function so it understands `mu4e` buffers as well; so put in your configuration:

```
(require 'gnus-dired)
;; make the `gnus-dired-mail-buffers' function also work on
;; message-mode derived modes, such as mu4e-compose-mode
(defun gnus-dired-mail-buffers ()
  "Return a list of active message buffers."
  (let (buffers)
    (save-current-buffer
      (dolist (buffer (buffer-list t))
(set-buffer buffer)
(when (and (derived-mode-p 'message-mode)
```

---

[3] Note, despite its name, `mu-cite` is a project unconnected to `mu`/`mu4e`

```
      (null message-sent-message-via))
    (push (buffer-name buffer) buffers))))
      (nreverse buffers)))

(setq gnus-dired-mail-mode 'mu4e-user-agent)
(add-hook 'dired-mode-hook 'turn-on-gnus-dired-mode)
```

Then, mark the file(s) in dired you would like to attach and press C-c RET C-a, and you'll be asked whether to attach them to an existing message, or create a new one.

Appendix B Example configurations

In this chapter, we show some example configurations. While it is very useful to see some working settings, we'd like to warn against blindly copying such things.

B.1 Minimal configuration

An (almost) minimal configuration for mu4e might look like this - as you see most is commented-out.

```
;; example configuration for mu4e

;; make sure mu4e is in your load-path
(require 'mu4e)

;; Only needed if your maildir is _not_ ~/Maildir
;;(setq mu4e-maildir "/home/user/Maildir")

;; these must start with a "/", and must exist
;; (i.e.. /home/user/Maildir/sent must exist)
;; you use e.g. 'mu mkdir' to make the Maildirs if they don't
;; already exist

;; below are the defaults; if they do not exist yet, mu4e offers to
;; create them. they can also functions; see their docstrings.
;; (setq mu4e-sent-folder   "/sent")
;; (setq mu4e-drafts-folder "/drafts")
;; (setq mu4e-trash-folder  "/trash")

;; smtp mail setting; these are the same that `gnus' uses.
(setq
   message-send-mail-function   'smtpmail-send-it
   smtpmail-default-smtp-server "smtp.example.com"
   smtpmail-smtp-server         "smtp.example.com"
   smtpmail-local-domain        "example.com")
```

B.2 Longer configuration

A somewhat longer configuration, showing some more things that you can customize.

```
;; example configuration for mu4e
(require 'mu4e)

;; path to our Maildir directory
(setq mu4e-maildir "/home/user/Maildir")

;; the next are relative to `mu4e-maildir'
;; instead of strings, they can be functions too, see
;; their docstring or the chapter 'Dynamic folders'
```

```
(setq mu4e-sent-folder    "/sent"
      mu4e-drafts-folder "/drafts"
      mu4e-trash-folder  "/trash")

;; the maildirs you use frequently; access them with 'j' ('jump')
(setq   mu4e-maildir-shortcuts
    '(("/archive"       . ?a)
      ("/inbox"         . ?i)
      ("/work"          . ?w)
      ("/sent"          . ?s)))

;; a  list of user's e-mail addresses
(setq mu4e-user-mail-address-list '("foo@bar.com" "cuux@example.com")

;; when you want to use some external command for text->html
;; conversion, e.g. the 'html2text' program
;; (setq mu4e-html2text-command "html2text")

;; the headers to show in the headers list -- a pair of a field
;; and its width, with `nil' meaning 'unlimited'
;; (better only use that for the last field.
;; These are the defaults:
(setq mu4e-headers-fields
     '( (:date          .  25)
        (:flags         .   6)
        (:from          .  22)
        (:subject       .  nil)))

;; program to get mail; alternatives are 'fetchmail', 'getmail'
;; isync or your own shellscript. called when 'U' is pressed in
;; main view.

;; If you get your mail without an explicit command,
;; use "true" for the command (this is the default)
(setq mu4e-get-mail-command "offlineimap")

;; general emacs mail settings; used when composing e-mail
;; the non-mu4e-* stuff is inherited from emacs/message-mode
(setq mu4e-reply-to-address "foo@bar.com"
      user-mail-address "foo@bar.com"
      user-full-name  "Foo X. Bar")
(setq mu4e-compose-signature
   "Foo X. Bar\nhttp://www.example.com\n")

;; smtp mail setting
(setq
   message-send-mail-function 'smtpmail-send-it
```

```
smtpmail-default-smtp-server "smtp.example.com"
smtpmail-smtp-server ""smtp.example.com"
smtpmail-local-domain "example.com"

;; if you need offline mode, set these -- and create the queue dir
;; with 'mu mkdir', i.e.. mu mkdir /home/user/Maildir/queue
smtpmail-queue-mail   nil
smtpmail-queue-dir   "/home/user/Maildir/queue/cur")

;; don't keep message buffers around
(setq message-kill-buffer-on-exit t)
```

B.3 Gmail configuration

Gmail is a popular e-mail provider; let's see how we can make it work with mu4e. Since we are using IMAP, you must enable that in the Gmail web interface (in the settings, under the "Forwarding and POP/IMAP"-tab).

Gmail users may also be interested in [Including related messages], page 30.

B.3.1 Setting up offlineimap

First of all, we need a program to get the e-mail from Gmail to our local machine; for this we use offlineimap; on Debian (and derivatives like Ubuntu), this is as easy as:

$ sudo apt-get install offlineimap

while on Fedora (and similar) you need:

$ sudo yum install offlineimap

Then, we can configure offlineimap by editing ~/.offlineimaprc:

```
[general]
accounts = Gmail
maxsyncaccounts = 3

[Account Gmail]
localrepository = Local
remoterepository = Remote

[Repository Local]
type = Maildir
localfolders = ~/Maildir

[Repository Remote]
type = IMAP
remotehost = imap.gmail.com
remoteuser = USERNAME@gmail.com
remotepass = PASSWORD
ssl = yes
maxconnections = 1
realdelete = no
```

Obviously, you need to replace USERNAME and PASSWORD with your actual Gmail username and password. After this, you should be able to download your mail:

```
$ offlineimap
OfflineIMAP 6.3.4
Copyright 2002-2011 John Goerzen & contributors.
Licensed under the GNU GPL v2+ (v2 or any later version).

Account sync Gmail:
***** Processing account Gmail
 Copying folder structure from IMAP to Maildir
 Establishing connection to imap.gmail.com:993.
Folder sync [Gmail]:
 Syncing INBOX: IMAP -> Maildir
 Syncing [Gmail]/All Mail: IMAP -> Maildir
 Syncing [Gmail]/Drafts: IMAP -> Maildir
 Syncing [Gmail]/Sent Mail: IMAP -> Maildir
 Syncing [Gmail]/Spam: IMAP -> Maildir
 Syncing [Gmail]/Starred: IMAP -> Maildir
 Syncing [Gmail]/Trash: IMAP -> Maildir
Account sync Gmail:
***** Finished processing account Gmail
```

We can now run mu to make sure things work:

```
$ mu index
mu: indexing messages under /home/foo/Maildir [/home/foo/.mu/xapian]
| processing mail; processed: 520; updated/new: 520, cleaned-up: 0
mu: elapsed: 3 second(s), ~ 173 msg/s
mu: cleaning up messages [/home/foo/.mu/xapian]
/ processing mail; processed: 520; updated/new: 0, cleaned-up: 0
mu: elapsed: 0 second(s)
```

We can run both the offlineimap and the mu index from within mu4e, but running it from the command line makes it a bit easier to troubleshoot as we are setting things up.

B.3.2 Settings

Next step: let's make a mu4e configuration for this:

```
(require 'mu4e)

;; default
;; (setq mu4e-maildir "~/Maildir")

(setq mu4e-drafts-folder "/[Gmail].Drafts")
(setq mu4e-sent-folder   "/[Gmail].Sent Mail")
(setq mu4e-trash-folder  "/[Gmail].Trash")

;; don't save message to Sent Messages, Gmail/IMAP takes care of this
(setq mu4e-sent-messages-behavior 'delete)
```

```
;; setup some handy shortcuts
;; you can quickly switch to your Inbox -- press ``ji''
;; then, when you want archive some messages, move them to
;; the 'All Mail' folder by pressing ``ma''.

(setq mu4e-maildir-shortcuts
    '( ("/INBOX"               . ?i)
       ("/[Gmail].Sent Mail"   . ?s)
       ("/[Gmail].Trash"       . ?t)
       ("/[Gmail].All Mail"    . ?a)))

;; allow for updating mail using 'U' in the main view:
(setq mu4e-get-mail-command "offlineimap")

;; something about ourselves
(setq
   user-mail-address "USERNAME@gmail.com"
   user-full-name  "Foo X. Bar"
   mu4e-compose-signature
    (concat
      "Foo X. Bar\n"
      "http://www.example.com\n"))

;; sending mail -- replace USERNAME with your gmail username
;; also, make sure the gnutls command line utils are installed
;; package 'gnutls-bin' in Debian/Ubuntu

(require 'smtpmail)
(setq message-send-mail-function 'smtpmail-send-it
   starttls-use-gnutls t
   smtpmail-starttls-credentials '(("smtp.gmail.com" 587 nil nil))
   smtpmail-auth-credentials
     '(("smtp.gmail.com" 587 "USERNAME@gmail.com" nil))
   smtpmail-default-smtp-server "smtp.gmail.com"
   smtpmail-smtp-server "smtp.gmail.com"
   smtpmail-smtp-service 587)

;; alternatively, for emacs-24 you can use:
;;(setq message-send-mail-function 'smtpmail-send-it
;;     smtpmail-stream-type 'starttls
;;     smtpmail-default-smtp-server "smtp.gmail.com"
;;     smtpmail-smtp-server "smtp.gmail.com"
;;     smtpmail-smtp-service 587)

;; don't keep message buffers around
(setq message-kill-buffer-on-exit t)
```

And that's it – put the above in your `~/.emacs`, change USERNAME etc. to your own, and restart emacs, and run *M-x mu4e*.

B.4 Some other useful settings

Finally, here are some more settings that are useful, but not enabled by default for various reasons.

```
;; use 'fancy' non-ascii characters in various places in mu4e
(setq mu4e-use-fancy-chars t)

;; save attachment to my desktop (this can also be a function)
(setq mu4e-attachment-dir "~/Desktop")

;; attempt to show images when viewing messages
(setq mu4e-view-show-images t)
```

Appendix C FAQ - Frequently Asked Questions

In this chapter we list a number of actual and anticipated questions and their answers.

C.1 General

1. *How can I quickly delete/move/trash a lot of messages?* You can select ('mark' in emacs-speak) the messages like you would select text in a buffer; the actions you then take (e.g., DEL for delete, m for move and t for trash) apply to all selected messages. You can also use functions like mu4e-headers-mark-thread (T), mu4e-headers-mark-subthread (t) to mark whole threads at the same time, and mu4e-headers-mark-pattern (%) to mark all messages matching a certain regular expression.

2. mu4e *seems to return a subset of all matches - how can I get all?* For speed reasons, mu4e returns only up to the value of the variable m4ue-search-result-limit (default: 500) matches. To show *all*, use *M-x mu4e-headers-toggle-full-search* (Q), or customize the variable mu4e-headers-full-search. This applies to all search commands.

3. *How can I get notifications when receiving mail?* There is mu4e-index-updated-hook, which gets triggered when the indexing process triggered sees an update (not just new mail though). To use this hook, put something like the following in your setup (assuming you have aplay and some soundfile, change as needed):

   ```
   (add-hook 'mu4e-index-updated-hook
     (defun new-mail-sound ()
       (shell-command "aplay ~/Sounds/boing.wav&")))
   ```

4. *It seems my headers-buffer is automatically updated when new messages are found during the indexing process – can I disable this somehow?* Yes - set mu4e-headers-auto-update to nil.

5. *I don't use* offlineimap, fetchmail *etc., I get my mail through my own mailserver. What should I use for* mu4e-get-mail-command? Use "true" (or don't do anything, it's the default). This makes getting mail a no-op, but the messages are still re-indexed.

6. *How can I re-index my messages without getting new mail?* Use *M-x mu4e-update-index*

7. *When I try to run* mu index *while* mu4e *is running I get errors like:*

   ```
   mu: mu_store_new_writable: xapian error
      'Unable to get write lock on ~/.mu/xapian: already locked
   ```

   What to do about this? You get this error because the underlying Xapian database is locked by some other process; it can be opened only once in read-write mode. There is not much mu4e can do about this, but if is another mu instance that is holding the lock, you can ask it to (gracefully) terminate:

   ```
   pkill -2 -u $UID mu # send SIGINT
   sleep 1
   mu index
   ```

   mu4e automatically restarts mu when it needs it. In practice, this seems to work quite well.

8. *I don't like the* Indexing... *messages that the indexing process gives me. Can I turn them off?.* Yes: set the variable mu4e-hide-index-messages to non-nil.

9. *Can I automatically apply the marks on messages when leaving the headers buffer?* Yes you can – see the documentation for the variable `mu4e-headers-leave-behavior`.

10. *Is there context-sensitive help available?* Yes - pressing H should take you to the right place in this manual.

11. *How can I set `mu4e` as the default e-mail client in `emacs`?* See Section A.1 [Setting the default emacs mail program], page 41.

12. *Can `mu4e` use some fancy Unicode characters instead of these boring plain-ASCII ones?* Glad you asked! Yes, if you set `mu4e-use-fancy-chars` to `t`, `mu4e` uses such fancy characters in a number of places.

13. *Can I start `mu4e` in the background?* Yes - if you provide a prefix-argument (`C-u`), `mu4e` starts, but does not show the main-window.

14. *Some IMAP-synchronization programs such as `mbsync` (but not `offlineimap`) don't like it when message files do not change their names when they are moved to different folders. Can `mu4e` somehow accomodate this?* Yes - you can set the variable `mu4e-change-filenames-when-moving` to non-nil.

15. `offlineimap` *uses IMAP's UTF-7 for encoding non-ascii folder names, while* mu *expects UTF-8 (so, e.g.* / [1] *becomes* /&MH4wijCCMEgwSg-). *How can display such folders correctly?* This is best solved by telling `offlineimap` to use UTF-8 instead – see `https://github.com/djcb/mu/issues/68#issuecomment-8598652`.

C.2 Reading messages

1. *How can I view attached images in my message view buffers?* See Section 5.4 [Viewing images inline], page 20.

2. *How can I word-wrap long lines in when viewing a message?* You can toggle between wrapped and non-wrapped states using `w`. If you want to do this automatically, invoke `longlines-mode` in your `mu4e-view-mode-hook`.

3. *What about hiding cited parts?* Toggle between hiding and showing of cited parts with `h`. If you want to hide parts automatically, call `mu4e-view-toggle-hide-cited` in your `mu4e-view-mode-hook`.

4. *How can I perform custom actions on messages and attachments?* See Chapter 10 [Actions], page 37.

5. *Does `mu4e` support crypto (i.e., decrypting messages and verifying signatures)?* Yes – if `mu` was built with `GMime` 2.6 or later, it is possible to do both (note, only PGP/MIME is supported). In the Chapter 3 [Main view], page 10 the support is indicated by a big letter `C` on the right hand side of the `mu4e` version. See [Decryption], page 21 and [Verifying signatures], page 21. For encryption and signing messages, see the Section C.3 [Writing messages], page 53.

6. *Does `mu4e` support including all related messages in a thread, like Gmail does?* Yes – see [Including related messages], page 30.

7. *There seem to be a lot of duplicate messages – how can I get rid of them?* See [Skipping duplicates], page 30.

---

[1] some Japanese characters, invisible in the UTF-8 version of this manual

8. *Some messages are almost unreadable in emacs - can I view them in an external web browser?* Indeed, airlines often send messages that heavily depend on html and are hard to digest inside emacs. Fortunately, there's an *action* (Section 10.3 [Adding an action in the message view], page 38) defined for this. Simply add to your configuration:

```
(add-to-list 'mu4e-view-actions
  '("ViewInBrowser" . mu4e-action-view-in-browser) t)
```

Now, when viewing such a difficult message, type aV, and the message opens inside a webbrowser. You can influence the browser with `browse-url-generic-program`.

C.3 Writing messages

1. *What's the deal with replies to messages I wrote myself?* Like many other mail-clients, mu4e treats replies to messages you wrote yourself as special – these message keep the same To: and Cc: as the original message. This is to ease the common case of following up to a message you wrote earlier.

2. *How can I automatically set the From:-address for a reply-message, based on some field in the original?* See Section 6.4 [Compose hooks], page 24.

3. *And what about customizable folders for draft messages, sent messages, trashed messages, based on e.g. the From: header?* See Chapter 9 [Dynamic folders], page 35.

4. *How can I automatically add some header to an outgoing message?* Once more, see Section 6.4 [Compose hooks], page 24.

5. *How can I influence the way the original message looks when replying or forwarding?* Since mu4e-compose-mode derives from `message-mode`, you can re-use many of the latter's facilities. See Info file `message`, node 'Insertion Variables'.

6. *How can I easily include attachments in the messages I write?* You can drag-and-drop from your desktop; alternatively, you can use dired – see Section A.8 [Attaching files with dired], page 43.

7. mu4e *seems to remove myself from the* Cc:-*list; how can I prevent that?* Set mu4e-compose-keep-self-cc to t in your configuration.

8. *How can I sign or encrypt messages?* You can do so using emacs' MIME-support – check the Attachments-menu while composing a message. Also see Section 6.5 [Signing and encrypting], page 25.

9. *Can I use BBDB with* mu4e*?* It should be possible, but there is no built-in support. Instead, we recommend using mu4e's Section 6.3 [Address autocompletion], page 23.

10. *After sending some messages, it seems the buffer for these messages stay around. How can I get rid of those?*

```
(setq message-kill-buffer-on-exit t)
```

11. *Sending big messages is slow and blocks emacs - what can I do about it?* For this, there's https://github.com/jwiegley/emacs-async (also available from the Emacs package repository); add the following snippet to your configuration:

```
(require 'smtpmail-async)
(setq
  send-mail-function 'async-smtpmail-send-it
  message-send-mail-function 'async-smtpmail-send-it)
```

With this, messages are sent using background emacs-instance.

A word of warning though, this tends to not be as reliable as sending the message in the normal, synchronous fashion, and people have reported silent failures, that is, message are not sent but there is no indication of that.

You can check the progress of the background by checking the *Messages*-buffer, which should show something like:

```
Delivering message to "William Shakespeare" <will@example.com>...
Mark set
Saving file /home/djcb/Maildir/sent/cur/20130706-044350-darklady:2,S...
Wrote /home/djcb/Maildir/sent/cur/20130706-044350-darklady:2,S
Sending...done
```

The first and final messages are the most important, and there may be considerable time between them, depending on the size of the message.

C.4 Known issues

Although they are not really *questions*, we end this chapter with a list of known issues and/or missing features in mu4e. Thus, users won't have to search in vain for things that are not there (yet), and the author can use it as a todo-list.

- *mu4e does not work well if the* emacs *language environment is not UTF-8*; so, if you problems with encodings, be sure to have (set-language-environment "UTF-8") in your ~/.emacs.

- *Thread handling is incomplete.* While threads are calculated and are visible in the headers buffer, you cannot collapse/open them.

- *The key-bindings are somewhat hard-coded.* That is, the main menu assumes the default key-bindings, as do the clicks-on-bookmarks.

- *The* emacs *front-end of the* notmuch *e-mail indexer conflicts with* mu4e. notmuch running in parallel with mu4e leads to

  ```
  error in process filter: mu4e-error-handler: Error 70: cannot read
  ~/Maildir/...
  ```

  when sending a reply to a new e-mail. This seems to be caused by notmuch changing the name of the original message file while mu4e is working in on it. To prevent this, deactivate notmuch in your Emacs setup.

- *The PDF-version of the manual does not show any of the non-ASCII characters* - this is because, sadly, texi2pdf documentation system does not support those. There is not much we can do about that.

For a more complete list, please refer to the issues-list in the github-repository.

Appendix D Tips and Tricks

D.1 Multiple accounts

Using mu4e with multiple email accounts is fairly easy. Although variables such as user-mail-address, mu4e-sent-folder, message-*, smtpmail-*, etc. typically only take one value, it is easy to change their values using mu4e-compose-pre-hook. The setup described here is one way of doing this (though certainly not the only way).

This setup assumes that you have multiple mail accounts under mu4e-maildir. As an example, we'll use ~/Maildir/Account1 and ~/Maildir/Account2, but the setup works just as well if mu4e-maildir points to something else.

First, you need to make sure that all variables that you wish to change based on user account are set to some initial value. So set up your environment with e.g., your main account:

```
(setq mu4e-sent-folder "/Account1/Saved Items"
      mu4e-drafts-folder "/Account1/Drafts"
      user-mail-address "my.address@account1.tld"
      smtpmail-default-smtp-server "smtp.account1.tld"
      smtpmail-local-domain "account1.tld"
      smtpmail-smtp-server "smtp.account1.tld"
      smtpmail-stream-type starttls
      smtpmail-smtp-service 25)
```

Then create a variable my-mu4e-account-alist, which should contain a list for each of your accounts. Each list should start with the account name, (which *must* be identical to the account's directory name under ~/Maildir), followed by (variable value) pairs:

```
(defvar my-mu4e-account-alist
  '(("Account1"
     (mu4e-sent-folder "/Account1/Saved Items")
     (mu4e-drafts-folder "/Account1/Drafts")
     (user-mail-address "my.address@account1.tld")
     (smtpmail-default-smtp-server "smtp.account1.tld")
     (smtpmail-local-domain "account1.tld")
     (smtpmail-smtp-server "smtp.account1.tld")
     (smtpmail-stream-type starttls)
     (smtpmail-smtp-service 25))
    ("Account2"
     (mu4e-sent-folder "/Account2/Saved Items")
     (mu4e-drafts-folder "/Account2/Drafts")
     (user-mail-address "my.address@account2.tld")
     (smtpmail-default-smtp-server "smtp.account2.tld")
     (smtpmail-local-domain "account2.tld")
     (smtpmail-smtp-server "smtp.account2.tld")
     (smtpmail-stream-type starttls)
     (smtpmail-smtp-service 587))))
```

You can put any variables you want in the account lists, just make sure that you put in *all* the variables that differ for each account. Variables that do not differ do not be included.

For example, if you use the same smtp server for both accounts, you don't need to include the smtp-related variables in `my-mu4e-account-alist`.

Now, the following function can be used to select an account and set the variables in `my-mu4e-account-alist` to the correct values:

```
(defun my-mu4e-set-account ()
  "Set the account for composing a message."
  (let* ((account
  (if mu4e-compose-parent-message
  (let ((maildir (mu4e-message-field mu4e-compose-parent-message :maildir)))
  (string-match "/\\(.*?\\)/" maildir)
  (match-string 1 maildir))
  (completing-read (format "Compose with account: (%s) "
  (mapconcat #'(lambda (var) (car var))
   my-mu4e-account-alist "/"))
  (mapcar #'(lambda (var) (car var)) my-mu4e-account-alist)
   nil t nil nil (caar my-mu4e-account-alist))))
  (account-vars (cdr (assoc account my-mu4e-account-alist))))
  (if account-vars
  (mapc #'(lambda (var)
  (set (car var) (cadr var)))
   account-vars)
  (error "No email account found"))))
```

This function then needs to be added to `mu4e-compose-pre-hook`:

```
(add-hook 'mu4e-compose-pre-hook 'my-mu4e-set-account)
```

This way, `my-mu4e-set-account` will be called every time you edit a message. If you compose a new message, it simply asks you for the account you wish to send the message from (TAB completion works). If you're replying or forwarding a message, or editing an existing draft, the account is chosen automatically, based on the first component of the maildir of the message being replied to, forwarded or edited (i.e., the directory under `~/Maildir`).

D.2 Refiling messages

By setting `mu4e-refile-folder` to a function, you can dynamically determine where messages are to be refiled. If you want to do this based on the subject of a message, you can use a function that matches the subject against a list of regexes in the following way. First, set up a variable `my-mu4e-subject-alist` containing regexes plus associated mail folders:

```
(defvar my-mu4e-subject-alist '(("kolloqui\\(um\\|a\\)" . "/Kolloquium")
                                ("Calls" . "/Calls")
                                ("Lehr" . "/Lehre")
                  ("webseite\\|homepage\\|website" . "/Webseite"))
    "List of subjects and their respective refile folders.")
```

Now you can use the following function to automatically refile messages based on their subject line:

```
(defun my-mu4e-refile-folder-function (msg)
  "Set the refile folder for MSG."
```

```
(let ((subject (mu4e-message-field msg :subject))
      (folder (or (cdar (member* subject my-mu4e-subject-alist
                                 :test #'(lambda (x y)
                                           (string-match (car y) x))))
                  "/General")))
  folder))
```

Note the `"/General"` folder: it is the default folder in case the subject does not match any of the regexes in `my-mu4e-subject-alist`.

In order to make this work, you'll of course need to set `mu4e-refile-folder` to this function:

```
(setq mu4e-refile-folder 'my-mu4e-refile-folder-function)
```

If you have multiple accounts, you can accommodate them as well:

```
(defun my-mu4e-refile-folder-function (msg)
  "Set the refile folder for MSG."
  (let ((maildir (mu4e-message-field msg :maildir))
        (subject (mu4e-message-field msg :subject))
        folder)
    (cond
     ((string-match "Account1" maildir)
      (setq folder (or (catch 'found
                         (dolist (mailing-list my-mu4e-mailing-lists)
                           (if (mu4e-message-contact-field-matches
                                msg :to (car mailing-list))
                               (throw 'found (cdr mailing-list)))))
                       "/Account1/General")))
     ((string-match "Gmail" maildir)
      (setq folder "/Gmail/All Mail"))
     ((string-match "Account2" maildir)
      (setq folder (or (cdar (member* subject my-mu4e-subject-alist
                                      :test #'(lambda (x y)
                                                (string-match
                                                 (car y) x))))
                       "/Account2/General"))))
    folder))
```

This function actually uses different methods to determine the refile folder, depending on the account: For *Account2*, it uses `my-mu4e-subject-alist`, for the *Gmail* account it simply uses the folder `"All Mail"`. For Account1, it uses another method: it files the message based on the mailing list to which it was sent. This requires another variable:

```
(defvar my-mu4e-mailing-lists
  '(("mu-discuss@googlegroups.com" . "/Account1/mu4e")
    ("pandoc-discuss@googlegroups.com" . "/Account1/Pandoc")
    ("auctex@gnu.org" . "/Account1/AUCTeX"))
  "List of mailing list addresses and folders where
   their messages are saved.")
```

D.3 Saving outgoing messages

Like `mu4e-refile-folder`, the variable `mu4e-sent-folder` can also be set to a function, in order to dynamically determine the save folder. One might, for example, wish to automatically put messages going to mailing lists into the trash (because you'll receive them back from the list anyway). If you have set up the variable `my-mu4e-mailing-lists` as mentioned, you can use the following function to determine a save folder:

```
(defun my-mu4e-sent-folder-function (msg)
  "Set the sent folder for the current message."
  (let ((from-address (message-field-value "From"))
        (to-address (message-field-value "To")))
    (cond
     ((string-match "my.address@account1.tld" from-address)
      (if (member* to-address my-mu4e-mailing-lists
                   :test #'(lambda (x y)
                             (string-match (car y) x)))
          "/Trash"
        "/Account1/Sent"))
     ((string-match "my.address@gmail.com" from-address)
      "/Gmail/Sent Mail")
     (t (mu4e-ask-maildir-check-exists "Save message to maildir: ")))))
```

Note that this function doesn't use (`mu4e-message-field msg :maildir`) to determine which account the message is being sent from. The reason is that that the function in `mu4e-sent-folder` is called when you send the message, but before mu4e has created the message struct from the compose buffer, so that `mu4e-message-field` cannot be used. Instead, the function uses `message-field-value`, which extracts the values of the headers in the compose buffer. This means that it is not possible to extract the account name from the message's maildir, so instead the from address is used to determine the account.

Again, the function shows three different possibilities: for the first account (`my.address@account1.tld`) it uses `my-mu4e-mailing-lists` again to determine if the message goes to a mailing list. If so, the message is put in the trash folder, if not, it is saved in `/Account1/Sent`. For the second (Gmail) account, sent mail is simply saved in the Sent Mail folder.

If the from address is not associated with Account1 or with the Gmail account, the function uses `mu4e-ask-maildir-check-exists` to ask the user for a maildir to save the message in.

D.4 Fancy characters and Inconsolata

When using 'fancy characters' (`mu4e-use-fancy-chars`) with the *Inconsolata*-font (and likely others as well), the display may be slightly off; the reason for this issue is that Inconsolata does not contain the glyphs for the 'fancy' arrows and the glyphs that are used as replacement are too high.

To fix this, you can use something like the following workaround (in your `.emacs`-file):

```
(if (equal window-system 'x)
    (progn
```

```
(set-fontset-font "fontset-default" 'unicode "Dejavu Sans Mono")
(set-face-font 'default "Inconsolata-10")))
```

D.5 Confirmation before sending

To protect yourself from sending messages too hastily, you can add a final confirmation, which you can of course make as elaborate as you wish.

```
(add-hook 'message-send-hook
  (lambda ()
    (unless (yes-or-no-p "Sure you want to send this?")
      (signal 'quit nil))))
```

Appendix E How it works

While perhaps not interesting for all users of mu4e, some curious souls may want to know how mu4e does its job.

E.1 High-level overview

At a high level, we can summarize the structure of the mu4e system using some ascii-art:

```
        +----------+
        | emacs    |
        |    +------+
        +----| mu4e | --> send mail (smtpmail)
             +------+
               | A
               V |   ---/ search, view, move mail
        +----------+     \
        |    mu    |
        +----------+
           |  A
          V   |
        +----------+
        | Maildir  |  <--- receive mail (fetchmail,
        +----------+                    offlineimap, ...)
```

In words:

- Your e-mail messages are stored in a Maildir-directory (typically, ~/Maildir and its subdirectories), and new mail comes in using tools like fetchmail, offlineimap, or through a local mail server.
- mu indexes these messages periodically, so you can quickly search for them. mu can run in a special **server**-mode, where it provides services to client software.
- mu4e, which runs inside emacs is such a client; it communicates with mu (in its **server**-mode to search for messages, and manipulate them.
- mu4e uses the facilities offered by emacs (the Gnus message editor and smtpmail) to send messages.

E.2 mu server

mu4e is based on the mu e-mail searching/indexer. The latter is a C-program; there are different ways to communicate with a client that is emacs-based.

One way to implement this, would be to call the mu command-line tool with some parameters and then parse the output. In fact, that was the first approach – mu4e would invoke e.g., mu find and process the output in emacs.

However, with this approach, we need to load the entire e-mail *Xapian* database (in which the message is stored) for each invocation. Wouldn't it be nicer to keep a running mu instance around? Indeed, it would - and thus, the **mu server** sub-command was born.

Running `mu` `server` starts a simple shell, in which you can give commands to `mu`, which then spits out the results/errors. `mu` `server` is not meant for humans, but it can be used manually, which is great for debugging.

E.3 Reading from the server

In the design, the next question was what format `mu` should use for its output for `mu4e` (`emacs`) to process. Some other programs use JSON here, but it seemed easier (and possibly, more efficient) just to talk to `emacs` in its native language: *s-expressions*, and interpret those using the `emacs`-function `read-from-string`. See Section E.4 [The message s-expression], page 61 for details on the format.

So, now let's look how we process the data from `mu` `server` in `emacs`. We'll leave out a lot of detail, `mu4e`-specifics, and look at a bit more generic approach.

The first thing to do is to create a process (for example, with `start-process`), and then register a filter function for it, which is invoked whenever the process has some data for us. Something like:

```
(let ((proc (start-process <arguments>)))
  (set-process-filter proc 'my-process-filter)
  (set-process-sentinel proc 'my-process-sentinel))
```

Note, the process sentinel is invoked when the process is terminated – so there you can clean things up. The function `my-process-filter` is a user-defined function that takes the process and the chunk of output as arguments; in `mu4e` it looks something like (pseudo-lisp):

```
(defun my-process-filter (proc str)
  ;; mu4e-buf: a global string variable to which data gets appended
  ;; as we receive it
  (setq mu4e-buf (concat mu4e-buf str))
  (when <we-have-received-a-full-expression>
      <eat-expression-from mu4e-buf>
      <evaluate-expression>))
```

`<evaluate-expression>` de-multiplexes the s-expression we got. For example, if the s-expression looks like an e-mail message header, it is processed by the header-handling function, which appends it to the header list. If the s-expression looks like an error message, it is reported to the user. And so on.

The language between frontend and backend is documented in the `mu-server` man-page. `mu4e` can log these communications; you can use *M-x mu4e-toggle-logging* to turn logging on and off, and you can view the log using *M-x mu4e-show-log* ($).

E.4 The message s-expression

A typical message s-expression looks something like the following:

```
(:docid 32461
 :from (("Nikola Tesla" . "niko@example.com"))
 :to (("Thomas Edison" . "tom@example.com"))
 :cc (("Rupert The Monkey" . "rupert@example.com"))
 :subject "RE: what about the 50K?"
 :date (20369 17624 0)
```

```
:size 4337
:message-id "C8233AB82D81EE81AF0114E4E74@123213.mail.example.com"
:path  "/home/tom/Maildir/INBOX/cur/133443243973_1.10027.atlas:2,S"
:maildir "/INBOX"
:priority normal
:flags (seen)
:parts ( (:index 1 :mime-type "text/plain" :size 12345 :attachment nil)
        (:index 2 :name "photo.jpg" :mime-type "image/jpeg"
         :size 147331 :attachment t)
        (:index 3 :name "book.pdf" :mime-type "application/pdf"
         :size 192220 :attachment t))
:references  ("C8384574032D81EE81AF0114E4E74@123213.mail.example.com"
"38203498230942D81EE81AF0114E4E74@123213.mail.example.com")
:in-reply-to "38203498230942D81EE81AF0114E4E74@123213.mail.example.com"
:body-txt "Hi Tom,

....
")
```

This s-expression forms a property list (`plist`), and we can get values from it using
`plist-get`; for example (`plist-get msg :subject`) would get you the message subject.
However, it's better to use the function `mu4e-message-field` to shield you from some
of the implementation details that are subject to change; and see the other convenience
functions in `mu4e-message.el`.

Some notes on the format:

- The address fields are *lists* of pairs (`name . email`), where `name` can be nil.

- The date is in format `emacs` uses (for example in `current-time`).[1]

- Attachments are a list of elements with fields `:index` (the number of the MIME-part),
  `:name` (the file name, if any), `:mime-type` (the MIME-type, if any) and `:size` (the size
  in bytes, if any).

- Messages in the Chapter 4 [Headers view], page 12 come from the database and do
  not have `:attachments`. `:body-txt` or `:body-html` fields. Message in the Chapter 5
  [Message view], page 17 use the actual message file, and do include these fields.

E.4.1 Example: ping-pong

As an example of the communication between `mu4e` and `mu`, let's look at the `ping-pong`-
sequence. When `mu4e` starts, it sends a command `ping` to the `mu server` backend, to
learn about its version. `mu server` then responds with a `pong` s-expression to provide this
information (this is implemented in `mu-cmd-server.c`).

We start this sequence when `mu4e` is invoked (when the program is started). It calls
`mu4e-proc-ping`, and registers a (lambda) function for `mu4e-proc-pong-func`, to handle
the response.

```
-> cmd:ping
<- (pong "mu" :version "x.x.x" :doccount 10000)
```

---

[1] Emacs 32-bit integers have only 29 bits available for the actual number; the other bits are use by emacs
for internal purposes. Therefore, we need to split `time_t` in two numbers.

When we receive such a `pong` (in `mu4e-proc.el`), the lambda function we registered is called, and it compares the version we got from the `pong` with the version we expected, and raises an error, if they differ.

Appendix F Logging and debugging

As explained in Appendix E [How it works], page 60, `mu4e` communicates with its backend (`mu server`) by sending commands and receiving responses (s-expressions).

For debugging purposes, it can be very useful to see this data. For this reason, `mu4e` can log all these messages. Note that the 'protocol' is documented to some extent in the `mu-server` manpage.

You can enable (and disable) logging with *M-x mu4e-toggle-logging*. The log-buffer is called `*mu4e-log*`, and in the Chapter 3 [Main view], page 10, Chapter 4 [Headers view], page 12 and Chapter 5 [Message view], page 17, there's a keybinding $ that takes you there. You can quit it by pressing q.

Logging can be a bit resource-intensive, so you may not want to leave it on all the time. By default, the log only maintains the most recent 1200 lines. `mu` itself keeps a log as well, you can find this it in `<MUHOME>/log/mu.log`, typically `~/.mu/log/mu.log`.

Appendix G GNU Free Documentation License

Version 1.2, November 2002

Copyright © 2000,2001,2002 Free Software Foundation, Inc.
51 Franklin St, Fifth Floor, Boston, MA 02110-1301, USA

Everyone is permitted to copy and distribute verbatim copies
of this license document, but changing it is not allowed.

0. PREAMBLE

The purpose of this License is to make a manual, textbook, or other functional and useful document *free* in the sense of freedom: to assure everyone the effective freedom to copy and redistribute it, with or without modifying it, either commercially or non-commercially. Secondarily, this License preserves for the author and publisher a way to get credit for their work, while not being considered responsible for modifications made by others.

This License is a kind of "copyleft", which means that derivative works of the document must themselves be free in the same sense. It complements the GNU General Public License, which is a copyleft license designed for free software.

We have designed this License in order to use it for manuals for free software, because free software needs free documentation: a free program should come with manuals providing the same freedoms that the software does. But this License is not limited to software manuals; it can be used for any textual work, regardless of subject matter or whether it is published as a printed book. We recommend this License principally for works whose purpose is instruction or reference.

1. APPLICABILITY AND DEFINITIONS

This License applies to any manual or other work, in any medium, that contains a notice placed by the copyright holder saying it can be distributed under the terms of this License. Such a notice grants a world-wide, royalty-free license, unlimited in duration, to use that work under the conditions stated herein. The "Document", below, refers to any such manual or work. Any member of the public is a licensee, and is addressed as "you". You accept the license if you copy, modify or distribute the work in a way requiring permission under copyright law.

A "Modified Version" of the Document means any work containing the Document or a portion of it, either copied verbatim, or with modifications and/or translated into another language.

A "Secondary Section" is a named appendix or a front-matter section of the Document that deals exclusively with the relationship of the publishers or authors of the Document to the Document's overall subject (or to related matters) and contains nothing that could fall directly within that overall subject. (Thus, if the Document is in part a textbook of mathematics, a Secondary Section may not explain any mathematics.) The relationship could be a matter of historical connection with the subject or with related matters, or of legal, commercial, philosophical, ethical or political position regarding them.

The "Invariant Sections" are certain Secondary Sections whose titles are designated, as being those of Invariant Sections, in the notice that says that the Document is released under this License. If a section does not fit the above definition of Secondary then it is not allowed to be designated as Invariant. The Document may contain zero Invariant Sections. If the Document does not identify any Invariant Sections then there are none.

The "Cover Texts" are certain short passages of text that are listed, as Front-Cover Texts or Back-Cover Texts, in the notice that says that the Document is released under this License. A Front-Cover Text may be at most 5 words, and a Back-Cover Text may be at most 25 words.

A "Transparent" copy of the Document means a machine-readable copy, represented in a format whose specification is available to the general public, that is suitable for revising the document straightforwardly with generic text editors or (for images composed of pixels) generic paint programs or (for drawings) some widely available drawing editor, and that is suitable for input to text formatters or for automatic translation to a variety of formats suitable for input to text formatters. A copy made in an otherwise Transparent file format whose markup, or absence of markup, has been arranged to thwart or discourage subsequent modification by readers is not Transparent. An image format is not Transparent if used for any substantial amount of text. A copy that is not "Transparent" is called "Opaque".

Examples of suitable formats for Transparent copies include plain ASCII without markup, Texinfo input format, LaTeX input format, SGML or XML using a publicly available DTD, and standard-conforming simple HTML, PostScript or PDF designed for human modification. Examples of transparent image formats include PNG, XCF and JPG. Opaque formats include proprietary formats that can be read and edited only by proprietary word processors, SGML or XML for which the DTD and/or processing tools are not generally available, and the machine-generated HTML, PostScript or PDF produced by some word processors for output purposes only.

The "Title Page" means, for a printed book, the title page itself, plus such following pages as are needed to hold, legibly, the material this License requires to appear in the title page. For works in formats which do not have any title page as such, "Title Page" means the text near the most prominent appearance of the work's title, preceding the beginning of the body of the text.

A section "Entitled XYZ" means a named subunit of the Document whose title either is precisely XYZ or contains XYZ in parentheses following text that translates XYZ in another language. (Here XYZ stands for a specific section name mentioned below, such as "Acknowledgements", "Dedications", "Endorsements", or "History".) To "Preserve the Title" of such a section when you modify the Document means that it remains a section "Entitled XYZ" according to this definition.

The Document may include Warranty Disclaimers next to the notice which states that this License applies to the Document. These Warranty Disclaimers are considered to be included by reference in this License, but only as regards disclaiming warranties: any other implication that these Warranty Disclaimers may have is void and has no effect on the meaning of this License.

2. VERBATIM COPYING

You may copy and distribute the Document in any medium, either commercially or noncommercially, provided that this License, the copyright notices, and the license notice saying this License applies to the Document are reproduced in all copies, and that you add no other conditions whatsoever to those of this License. You may not use technical measures to obstruct or control the reading or further copying of the copies you make or distribute. However, you may accept compensation in exchange for copies. If you distribute a large enough number of copies you must also follow the conditions in section 3.

You may also lend copies, under the same conditions stated above, and you may publicly display copies.

3. COPYING IN QUANTITY

If you publish printed copies (or copies in media that commonly have printed covers) of the Document, numbering more than 100, and the Document's license notice requires Cover Texts, you must enclose the copies in covers that carry, clearly and legibly, all these Cover Texts: Front-Cover Texts on the front cover, and Back-Cover Texts on the back cover. Both covers must also clearly and legibly identify you as the publisher of these copies. The front cover must present the full title with all words of the title equally prominent and visible. You may add other material on the covers in addition. Copying with changes limited to the covers, as long as they preserve the title of the Document and satisfy these conditions, can be treated as verbatim copying in other respects.

If the required texts for either cover are too voluminous to fit legibly, you should put the first ones listed (as many as fit reasonably) on the actual cover, and continue the rest onto adjacent pages.

If you publish or distribute Opaque copies of the Document numbering more than 100, you must either include a machine-readable Transparent copy along with each Opaque copy, or state in or with each Opaque copy a computer-network location from which the general network-using public has access to download using public-standard network protocols a complete Transparent copy of the Document, free of added material. If you use the latter option, you must take reasonably prudent steps, when you begin distribution of Opaque copies in quantity, to ensure that this Transparent copy will remain thus accessible at the stated location until at least one year after the last time you distribute an Opaque copy (directly or through your agents or retailers) of that edition to the public.

It is requested, but not required, that you contact the authors of the Document well before redistributing any large number of copies, to give them a chance to provide you with an updated version of the Document.

4. MODIFICATIONS

You may copy and distribute a Modified Version of the Document under the conditions of sections 2 and 3 above, provided that you release the Modified Version under precisely this License, with the Modified Version filling the role of the Document, thus licensing distribution and modification of the Modified Version to whoever possesses a copy of it. In addition, you must do these things in the Modified Version:

A. Use in the Title Page (and on the covers, if any) a title distinct from that of the Document, and from those of previous versions (which should, if there were any, be listed in the History section of the Document). You may use the same title as a previous version if the original publisher of that version gives permission.

B. List on the Title Page, as authors, one or more persons or entities responsible for authorship of the modifications in the Modified Version, together with at least five of the principal authors of the Document (all of its principal authors, if it has fewer than five), unless they release you from this requirement.

C. State on the Title page the name of the publisher of the Modified Version, as the publisher.

D. Preserve all the copyright notices of the Document.

E. Add an appropriate copyright notice for your modifications adjacent to the other copyright notices.

F. Include, immediately after the copyright notices, a license notice giving the public permission to use the Modified Version under the terms of this License, in the form shown in the Addendum below.

G. Preserve in that license notice the full lists of Invariant Sections and required Cover Texts given in the Document's license notice.

H. Include an unaltered copy of this License.

I. Preserve the section Entitled "History", Preserve its Title, and add to it an item stating at least the title, year, new authors, and publisher of the Modified Version as given on the Title Page. If there is no section Entitled "History" in the Document, create one stating the title, year, authors, and publisher of the Document as given on its Title Page, then add an item describing the Modified Version as stated in the previous sentence.

J. Preserve the network location, if any, given in the Document for public access to a Transparent copy of the Document, and likewise the network locations given in the Document for previous versions it was based on. These may be placed in the "History" section. You may omit a network location for a work that was published at least four years before the Document itself, or if the original publisher of the version it refers to gives permission.

K. For any section Entitled "Acknowledgements" or "Dedications", Preserve the Title of the section, and preserve in the section all the substance and tone of each of the contributor acknowledgements and/or dedications given therein.

L. Preserve all the Invariant Sections of the Document, unaltered in their text and in their titles. Section numbers or the equivalent are not considered part of the section titles.

M. Delete any section Entitled "Endorsements". Such a section may not be included in the Modified Version.

N. Do not retitle any existing section to be Entitled "Endorsements" or to conflict in title with any Invariant Section.

O. Preserve any Warranty Disclaimers.

If the Modified Version includes new front-matter sections or appendices that qualify as Secondary Sections and contain no material copied from the Document, you may at your option designate some or all of these sections as invariant. To do this, add their titles to the list of Invariant Sections in the Modified Version's license notice. These titles must be distinct from any other section titles.

You may add a section Entitled "Endorsements", provided it contains nothing but endorsements of your Modified Version by various parties—for example, statements of peer review or that the text has been approved by an organization as the authoritative definition of a standard.

You may add a passage of up to five words as a Front-Cover Text, and a passage of up to 25 words as a Back-Cover Text, to the end of the list of Cover Texts in the Modified Version. Only one passage of Front-Cover Text and one of Back-Cover Text may be added by (or through arrangements made by) any one entity. If the Document already includes a cover text for the same cover, previously added by you or by arrangement made by the same entity you are acting on behalf of, you may not add another; but you may replace the old one, on explicit permission from the previous publisher that added the old one.

The author(s) and publisher(s) of the Document do not by this License give permission to use their names for publicity for or to assert or imply endorsement of any Modified Version.

5. COMBINING DOCUMENTS

You may combine the Document with other documents released under this License, under the terms defined in section 4 above for modified versions, provided that you include in the combination all of the Invariant Sections of all of the original documents, unmodified, and list them all as Invariant Sections of your combined work in its license notice, and that you preserve all their Warranty Disclaimers.

The combined work need only contain one copy of this License, and multiple identical Invariant Sections may be replaced with a single copy. If there are multiple Invariant Sections with the same name but different contents, make the title of each such section unique by adding at the end of it, in parentheses, the name of the original author or publisher of that section if known, or else a unique number. Make the same adjustment to the section titles in the list of Invariant Sections in the license notice of the combined work.

In the combination, you must combine any sections Entitled "History" in the various original documents, forming one section Entitled "History"; likewise combine any sections Entitled "Acknowledgements", and any sections Entitled "Dedications". You must delete all sections Entitled "Endorsements."

6. COLLECTIONS OF DOCUMENTS

You may make a collection consisting of the Document and other documents released under this License, and replace the individual copies of this License in the various documents with a single copy that is included in the collection, provided that you follow the rules of this License for verbatim copying of each of the documents in all other respects.

You may extract a single document from such a collection, and distribute it individually under this License, provided you insert a copy of this License into the extracted document, and follow this License in all other respects regarding verbatim copying of that document.

7. AGGREGATION WITH INDEPENDENT WORKS

A compilation of the Document or its derivatives with other separate and independent documents or works, in or on a volume of a storage or distribution medium, is called an "aggregate" if the copyright resulting from the compilation is not used to limit the legal rights of the compilation's users beyond what the individual works permit. When the Document is included in an aggregate, this License does not apply to the other works in the aggregate which are not themselves derivative works of the Document.

If the Cover Text requirement of section 3 is applicable to these copies of the Document, then if the Document is less than one half of the entire aggregate, the Document's Cover Texts may be placed on covers that bracket the Document within the aggregate, or the electronic equivalent of covers if the Document is in electronic form. Otherwise they must appear on printed covers that bracket the whole aggregate.

8. TRANSLATION

Translation is considered a kind of modification, so you may distribute translations of the Document under the terms of section 4. Replacing Invariant Sections with translations requires special permission from their copyright holders, but you may include translations of some or all Invariant Sections in addition to the original versions of these Invariant Sections. You may include a translation of this License, and all the license notices in the Document, and any Warranty Disclaimers, provided that you also include the original English version of this License and the original versions of those notices and disclaimers. In case of a disagreement between the translation and the original version of this License or a notice or disclaimer, the original version will prevail.

If a section in the Document is Entitled "Acknowledgements", "Dedications", or "History", the requirement (section 4) to Preserve its Title (section 1) will typically require changing the actual title.

9. TERMINATION

You may not copy, modify, sublicense, or distribute the Document except as expressly provided for under this License. Any other attempt to copy, modify, sublicense or distribute the Document is void, and will automatically terminate your rights under this License. However, parties who have received copies, or rights, from you under this License will not have their licenses terminated so long as such parties remain in full compliance.

10. FUTURE REVISIONS OF THIS LICENSE

The Free Software Foundation may publish new, revised versions of the GNU Free Documentation License from time to time. Such new versions will be similar in spirit to the present version, but may differ in detail to address new problems or concerns. See http://www.gnu.org/copyleft/.

Each version of the License is given a distinguishing version number. If the Document specifies that a particular numbered version of this License "or any later version" applies to it, you have the option of following the terms and conditions either of that specified version or of any later version that has been published (not as a draft) by the Free Software Foundation. If the Document does not specify a version number of this License, you may choose any version ever published (not as a draft) by the Free Software Foundation.

ADDENDUM: How to use this License for your documents

To use this License in a document you have written, include a copy of the License in the document and put the following copyright and license notices just after the title page:

```
Copyright (C)  year  your name.
Permission is granted to copy, distribute and/or modify this document
under the terms of the GNU Free Documentation License, Version 1.2
or any later version published by the Free Software Foundation;
with no Invariant Sections, no Front-Cover Texts, and no Back-Cover
Texts.  A copy of the license is included in the section entitled ``GNU
Free Documentation License''.
```

If you have Invariant Sections, Front-Cover Texts and Back-Cover Texts, replace the "with...Texts." line with this:

```
with the Invariant Sections being list their titles, with
the Front-Cover Texts being list, and with the Back-Cover Texts
being list.
```

If you have Invariant Sections without Cover Texts, or some other combination of the three, merge those two alternatives to suit the situation.

If your document contains nontrivial examples of program code, we recommend releasing these examples in parallel under your choice of free software license, such as the GNU General Public License, to permit their use in free software.